Basic Statistical Techniques for Medical and Other Professionals

Basic Statistical Techniques for Medical and Other Professionals

A Course in Statistics to Assist in Interpreting Numerical Data

Dr. David J. Smith
Foreword by Sam Samuel

Routledge
Taylor & Francis Group

A PRODUCTIVITY PRESS BOOK

First Published 2022
by Routledge
605 Third Avenue, New York, NY 10158

and by Routledge
2 Park Square, Milton Park, Abingdon, Oxon, OX14 4RN

Routledge is an imprint of the Taylor & Francis Group, an Informa business

Library of Congress Cataloging-in-Publication Data
A catalog record for this title has been requested

ISBN: 978-1-032-11495-8 (hbk)
ISBN: 978-1-032-11494-1 (pbk)
ISBN: 978-1-003-22013-8 (ebk)

DOI: 10.4324/9781003220138

Typeset in Garamond
by MPS Limited, Dehradun

Contents

Foreword

I have had the pleasure of knowing David Smith for nearly 40 years. He has run a successful consultancy business involving Reliability and Risk throughout that time, which is heavily dependent on statistical analysis.

The idea for this book arose because a relative, having attained a degree in Neuroscience, was undertaking an MSc to further her career, which has a significant statistical element. David initially produced training material to provide the basic skills required. During the course, the material was continually updated, to the point where he considered the exercises and slides worth converting into a textbook. My conclusion is that it was the right decision.

Although the book is aimed specifically at members of the medical fraternity (including students), it concentrates on practical analysis methods rather than the underlying theory. It is a compromise between simple practical methods and unnecessary mathematics and should be equally valuable to anyone taking a high-level degree course that involves knowledge of basic statistical techniques. It is also aimed at those already qualified and in practice in most professions where numerical data is involved.

Sam Samuel

Preface

The purpose of this book is to provide members of the medical and other professions, including scientists and engineers, with a basic understanding of statistics and probability. It does not seek to confuse the reader with in-depth mathematics but provides basic methods for interpreting data and making inferences. Although the worked examples have a medical flavour the principles apply to the analysis of any numerical data.

The information in this book is for guidance and is not intended to give specific instructions to professionals, with whom the sole responsibility for patient/client safety lies. No responsibility is assumed by the author/publisher for any consequences arising from actions taken on the basis of the material herein.

Acknowledgements

My sincere thanks to Sam Samuel, without whose patience and help I would not have completed this little book. He has been a life-long colleague, associate and friend and is a highly respected consultant in the specialised field of quantified risk assessment. His very major contributions to the text and material in this book have been invaluable.

About the Author

Dr David J Smith BSc, PhD, FIET, FCGI, HonFSaRS, MIGEM, is the Proprietor of Technis Consultancy. He has written numerous books on Reliability, Risk and Statistics during the last 40 years. He has also written a range of articles (and a book) on the statistics of helicopter safety. David is the past President of the UK Safety and Reliability Society.

By the same author:
Reliability Maintainability and Risk, 9th Edition, Smith DJ, Elsevier ISBN 9780081020104.
The Safety Critical Systems Handbook 5th edition, Smith DJ and Simpson KGL, (Elsevier) ISBN 9780128207000.

Introduction

We are surrounded by numerical data in every walk of life. Medical professionals record blood chemistries, blood pressures, blood sugars, response times, optical measurements, pain and quality-of-life-related metrics and so on. The list is endless. Engineers and scientists practice only by virtue of recording and making judgements from numbers, which record the physical parameters that describe our world. Every profession involves some aspect of counting something, be it money, persons, times, market responses, patterns of criminality and so on.

There are numerous excellent textbooks on the subject of statistics that explain the mathematics in detail and provide a whole range of sophisticated techniques for analysing numerical data. Those embarking on a career of scientific research (medical or otherwise) need to study the subject at that level. However, the vast majority of professionals do not require that depth, and this book is intended to satisfy the need for a relatively short course which explains and illustrates simple, but nevertheless powerful, techniques for making important judgements that would otherwise not be arrived at.

I may be accused of oversimplifying the subject but I am convinced that the techniques described here are more than adequate for the vast majority of people.

Statistics has been called, amongst other things, an inexact science and is often regarded with suspicion by the uninitiated. The subject is surrounded by misunderstanding and, furthermore, the distinction between probability and statistics is not always clear. We shall, therefore, begin by looking at the meaning of the two terms.

Statistics is simply the science of interpreting numerical data. If, in a certain brand of matches, 99.9% fall within the range 4.8 cm, plus or minus 0.3 cm, then it could be inferred that a match 5.25 cm in length chosen, at random, from a number of brands is unlikely to belong to the brand in question. The purpose of statistics is to determine, subject to various assumptions, exactly how likely it is that our inference is correct (e.g., 90% sure; 99% sure).

"99% sure" is another way of saying that we expect our inference to be proved wrong approximately only one time in 100. Statistics is not, therefore, an inexact science, but the science of being exact about the degree of uncertainty in a statement which was based on numerical data. Statistics, of course, cannot lie but those who use them can paint an untrue picture by ignoring inconvenient measurements, by making unjustified assumptions, and by omitting to tell the WHOLE truth. By the end of this book, you will be better equipped to test for yourself the validity of statistically based statements.

Now let us examine the meaning of the word Probability. The probability of an event can be thought of as the ratio of the number of occurrences of that event (i.e., successes) to the total number of items of data, providing that the experiment continues indefinitely. We often wish to determine the probability of some complex situation occurring where the interaction of a number of events is needed, each of which whose probabilities are known. Probability theory involves manipulating certain rules in order to determine the likelihood of such outcomes.

Since, as we saw in an earlier paragraph, statistics involves stating the probability of some numerical inference being correct then, clearly, there is a link between statistics and probability theory.

Since an estimate of probability is given by the ratio of the number of occurrences (i.e., successes) of an event to the number of items, it can be arrived at in two ways.

If it is established that, for UK males, 10% exceed 6 feet in height then, from that information, the probability that a person, chosen at random from a group of UK males, exceeds 6 feet in height is indeed 10% (i.e., 0.1). This is an empirical statement of probability being derived from observed data.

If, on the other hand, we wish to know the probability of drawing a heart from a deck of playing cards it can be argued that since 25% of the cards are hearts then a card, drawn at random, has a 0.25 probability of being a heart. This is a probability statement based on prior knowledge of the population and NOT by experimentally drawing cards repeatedly in order to determine the proportion of hearts. We call this an à priori statement of probability.

In both cases, the concept of probability is derived from a proportion of successes and is thus a dimensionless quantity (that is to say it takes no units). A probability can take any value between zero (impossibility) and unity (certainty).

Spreadsheets are an excellent and time-saving way of carrying out statistical calculations. Those readers who are not regular spreadsheet users would be wise to study Appendix 1 carefully. It provides a "refresher" in manipulating numbers quickly. Many of the exercises throughout this book use the spreadsheet format and the reader is encouraged to use one when solving them.

Now, read on!

Chapter 1

Why One Needs Statistical Techniques

We are frequently faced with numerical information (known as data) which, in its raw form, gives no clear picture of the trends that it might contain. The following simple example illustrates the point, as well as introducing two basic concepts.

The blood sugar levels in mmol/L (millimoles per litre), recorded by a diabetic patient over a period of 10 days, were

7.5, 7.2, 8, 6, 5.6, 5, 7, 5.5, 5, 5.5

The average can easily be calculated as 6.2 although it may not be obvious merely by glancing at the row of numbers. Furthermore, the graph (Figure 1.1) shows that the trend is, to some extent, downwards.

Thus, albeit at a very basic level, two important ideas have emerged:

An **average** (we use the word MEAN in statistical work)
and
Providing a **visual plot** of the trend (a linear GRAPH)

DOI: 10.4324/9781003220138-1

1

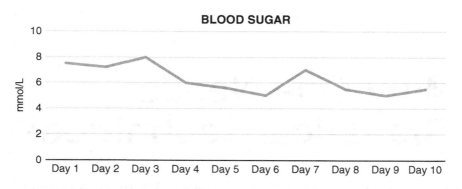

Figure 1.1   Simple graph of a variable against time

Both can be extremely helpful.

Notice the use of the word "linear" in the brackets. Linear means that the distances separating the days along the x axis and the distances separating the blood sugar levels, ascending the y axis, are equal. This will not always be the case in more advanced uses of the "graph" technique (Chapter 5). We will return to that subject later.

Taking the concept of a MEAN further we might add, providing the spread of values on either side is similar, that we can make statements such as "I am 50% sure (in statistics we use the word "confident") that any value, taken at random, is greater than 6.2." That idea will be developed in Chapter 4 to make more sophisticated inferences, such as being 90% confident of exceeding some stated value.

In order to do that, we need to think about the variability of the data. In other words, the spread ("distribution") of values between the two extreme values which were the two sugar levels of 5 and 8 mmol/L in the earlier example. Chapter 4 will develop this idea further and later Chapters will provide techniques for making useful comparisons between different sets of data.

Variables and Attributes

At this point, it will be useful to distinguish between Variables and Attributes.

Variables are a measure of an item of interest (e.g., size, weight, time, blood sugar, systolic blood pressure) and can take any value in a given range (e.g., feet, ounces, minutes, mmol/L [blood sugar], mmHg [blood pressure].

Attributes, however, are binary and describe some state that either applies or does not apply. In a deck of cards any one card can be a heart (or not). A person can be either alive or dead. One cannot be a bit dead, only alive or dead. Thus, attributes are measured in numbers of items having that attribute (or state) and have no units.

Spread of a Variable

The blood sugar readings illustrated in Figure 1.1 are distributed (spread) around the mean value of 6.2 mmol/L. The "tightness" or "looseness" of that spread is important since it describes the consistency (or otherwise) of the measurement in question. In the blood sugar example, a consistent reading, shown by a tighter scatter, would be desirable. Statistics will provide a way of describing how consistent the readings are.

Correlation

Another problem, which statistics will help to unravel, is the need to establish if some variable is related to another. One example would be diastolic blood pressure and systolic blood pressure. One does not precisely dictate the value of the other, but it might be credible to assume that a change in one might be followed by a similar change in the other. Again, statistics will provide a more precise way of quantifying the strength of the interaction. This particular suggestion will be dealt with in Chapter 9 where we will test the strength of association between the two.

Taking Samples

Much of statistical analysis consists of drawing conclusions from a set of data or of comparing two or more sets of data. The data in question is nearly always a sample drawn from a wider (larger) population. It is therefore important to think about whether the data that one has gathered is, indeed, a representative sample of the population of interest.

Random sampling is the ideal way of collecting data from a population. To achieve this, each item in the population needs to have an equal chance of being chosen for the sample. If, for example, the total annual population of stroke sufferers is 115,000 (UK) and in conducting a comparison between two alternative treatment regimes, two sample groups of 50 patients were selected, there is always the possibility that the sample did not represent the population as a whole.

It is important to take an unbiased random sample from which to draw conclusions. Thus, when we selected the two samples of 50 patients from the total population of 115,000, there was the possibility that the researcher might have selected 50% men and 50% women for the sample even though the stroke population might contain say 67% men and 33% women. The sample would not then accurately reflect some gender-related factor.

Gender is, of course, not the only relevant factor since age, ethnicity, lifestyle, diet and occupation might all be argued to be relevant. The problem with random sampling is that it requires a complete knowledge of the population before selecting the sample.

Very Large and Very Small Numbers

In many fields, data involves numbers in the hundreds of thousands and greater and deals with factors such as one in a hundred thousand and so on. "One in a hundred thousand" is a cumbersome way of expressing data and for those not familiar with the convention of expressing numbers as positive and negative powers of ten, Chapter 9 provides a thorough grounding.

Chapter 2

Probability and Its Rules

Empirical versus À Priori

An estimate of the probability of an event is given by the ratio of the number of occurrences (i.e., successes) of that event to the number of items of data, and this can be arrived at in two ways. Revisiting what was mentioned in the introduction:

If it is established that, in the UK (population 67 Million), there are 4.7 Million diabetics then, from that information, the probability that a person, chosen at random, is diabetic is 7% (i.e., 4.7/67). This is a probability statement based on prior knowledge of the population and NOT by experimentally observing a sample. We call this an **à priori** statement of probability.

If, on the other hand, we estimate the probability by observing that, in a large medical practice covering 8,000 patients, there are 480 diabetics, then we might infer that the probability of being diabetic is 6% (i.e., 480/8,000). This is an **empirical** statement of probability being derived from sample data.

DOI: 10.4324/9781003220138-2

7

In both cases, the concept of probability is derived from a proportion of successes and is therefore a dimensionless quantity (that is to say it takes no units). A probability can take any value between zero (impossibility) and unity (certainty).

Combining Probabilities

Assume that the probability of a person, selected at random, exceeding 5'6" in height (Event A) is 0.8 and that the probability of a person attaining the age of 65 (Event B) is 0.7. Figure 2.1 represents this state of affairs by what is known as a Venn diagram. The fact that the circles overlap represents the fact that the two events are not mutually exclusive – clearly it is possible to exceed a height of 5'6" and, also, to attain the age of 65 years. The area outside the circles represents all possibilities which are neither event A nor event B.

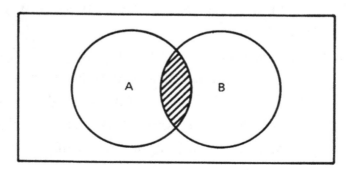

Figure 2.1   Overlapping events

The shaded (overlapping) portion of the diagram re-presents the possibility that both event A and event B will take place and the reader will doubtlessly recognise this as the situation to which the Multiplication rule applies.

If Pa (the probability of event A occurring) is 0.8
If Pb (the probability of event B occurring) is 0.7

Then the probability that both event A **and** event B will occur is

$$\text{Pab} = \text{Pa} \times \text{Pb} = 0.8 \times 0.7 = 0.56$$

In other words, 0.8 proportion of the 0.7 proportion of the total number of trials will result in both A and B together. The rule can be extended to calculate the probability that all of n events will occur together.

$$\text{Pan} = \text{Pa} \times \text{Pb} \times \text{Pc} \ldots \ldots \ldots \times \text{Pn}$$

The above equations assume independence between the events Pa, Pb and so on. Consider the following case. The probability of a new-born cot death is approximately one in 8,000 births. Thus, the probability of two babies dying in the same family might be inferred to be one in 8,000 x 8,000 = one in 64 Million. This assumes that if a child in a family dies of infant death syndrome, then it is no more and no less likely that a subsequent child will die in that way. However, there are reasons why the two events might not be independent as, for example, a genetic factor, or an environmental factor unique to that family.

A different calculation involves the probability of either A **or** B **or** both (in other words at least one) of the events occurring. The answer, in this case, is

$$P(a \ or \ b \ or \ both) = Pa + Pb - (Pa \ x \ Pb)$$
(More easily remembered as SUM − PRODUCT)

Examining this in terms of ratios, 0.8 of events will turn out to be A and, of these, some will also be B. Furthermore 0.7 of events will turn out to be B and, of these, some will also be A. Those events which are both A and B have, therefore, been included twice and must be allowed for by deducting Pa x Pb, namely the proportion which are indeed both A and B.

Another way of arriving at the same answer is to argue that:

1−Pa is the probability of event A NOT occurring
1−Pb is the probability of event B NOT occurring
Thus, $(1-Pa) \times (1-Pb)$ is the probability that an event will be neither A nor B.
Therefore $1 - (1-Pa) \times (1-Pb)$ is the probability of either A or B or both.
However: $1 - (1-Pa) \times (1-Pb) =$ **Pa + Pb − (Pa x Pb)** as above.

It follows that for n events the Probability of one or more occurring is

$$1 - (1 - Pa)\ (1 - Pb)\ (1 - Pc)\ ......\ (1 - Pn).$$

Mutually Exclusive Events

Let us now examine the case of mutually exclusive events. Consider a deck of cards and let event A be the drawing of a heart and event B the drawing of a black card. This time there is no overlap since a card cannot satisfy both outcomes. This is illustrated in Figure 2.2.

The previous theorem does not apply here since the probability of both event A **and** event B taking place is zero. It follows that the probability of drawing either a heart or a black card (event A or event B) is

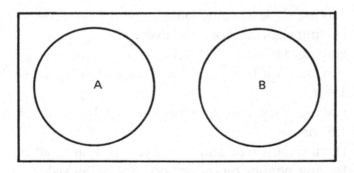

Figure 2.2 Mutually exclusive events

$$Pa + Pb$$

and for multiple events, the probability of observing either event A, or event B, or event C, or event D, etc., is

$$Pa + Pb + Pc + Pd \text{ etc}$$

In the playing card example Pa = 0.25 Pb = 0.5 and therefore the probability of drawing either a heart or a black card is 0.25 + 0.5 = 0.75.

Exercise 1   Manipulating Probabilities

Assume that:

Having dark hair is 80% likely.
Having blue eyes is 25% likely.
Being bald is 5% likely.
Being taller than 1.8 m is 50% likely.

What is the probability of the following:

1. Having dark hair and blue eyes?
2. Having dark hair or blue eyes?
3. Having dark hair and being bald?
4. Having dark hair and blue eyes and being taller than 1.8 m?
5. Having dark hair or blue eyes or being taller than 1.8 m?
6. Having dark hair and blue eyes, or being bald?
7. Having neither blue eyes nor not being bald?

Conditional Probabilities

The following example illustrates a state of affairs in which the probability of an event is conditional on the probability of some other factor. Assume that a person, at random, has a 1% probability of suffering from a particular form of cancer (in other words 1% of the population are known to suffer). Assume, as is often the case, that there is a test to determine if the cancer is present but that the test is not perfect. Like many tests, it may return a false positive despite the subject not suffering from the condition and it might also return a false negative in that it fails to detect a real case. The picture is summarised in the following table with the assumption that there is a 10% chance of a false result (be it positive or negative).

	Patient Has Cancer (1%)	Patient Has No Cancer (99%)
POSITIVE TEST (Suggests cancer is present)	90% True Positive	10% False Positive
NEGATIVE TEST (Suggests cancer is not present)	10% False Negative	90% True Negative

Given a Positive Test

If one is tested at random (given the 1% chance of having the disease), the probability of being a true positive is 90% of 1% = **0.9%.**

By a similar argument, the probability of a false positive is 10% of (100% − 1%) = 10% of 99% = **9.9%**.

Therefore, the probability of seeing any positive result is 0.9% + 9.9% = **10.8%**

Since probability is the ratio of an event in question to the number of trials then, we can argue that the probability of having cancer (***given that one receives a positive result***) is

> (Probability of a true positive)/(Probability of any positive)
> = 0.9%/10.8% = **8.3%**

Given a Negative Test

If one is tested at random (given the 99% chance of not having the disease), the probability of being a true negative is 90% of 99% = **89.1%.**

By a similar argument, the probability of a false negative is 10% of (100%–99%) = 10% of 1% = **0.1%**.

Therefore, the probability of seeing any negative result is 89.1% + 0.1% = **89.2%**

Since probability is the ratio of an event in question to the number of trials then, we can argue that the probability of not having cancer (***given that one receives a negative result***) is

(Probability of a true negative)/(Probability of any negative)
= 89.1%/89.2% = **99.9%**

Comparison of the Two Outcomes

In the case of testing positive the chances of a result, despite no illness, are swamped by the false positives of 99% of the population, resulting in a small probability (8.3%) of actually having the illness. On the other hand, the negative result is a more reliable indicator due to the small contribution of false negatives from those suffering from the condition. This all rests on the proportion of the population suffering from the condition. In the case of 50%, then both positive and negative tests would yield the same confidence of being correct.

The reader may care to populate the above examples with alternative data relating to other conditions, the data required for the study being:

- Percentage of the population suffering from the condition
- Percentage of false-positive tests (despite no condition existing)
- Percentage of false-negative tests (despite the condition existing)

In the above examples, for simplicity, the same values for false positive and false negative results were used. In practice, it is frequently the case that they are not the same and are somewhat unbalanced.

The above equations are an example of Bayes Theorem. In summary:

Probability of Cancer (given a positive result)

$$= \textit{Probability of Cancer} \times \frac{\textit{Probability of a positive result if Cancer is present}}{\textit{Probability of any positive result}}$$

The equivalent equation applies to the negative result.

Appendix 1 provides a spreadsheet aid to calculating the results. It can be used to evaluate the data from different tests.

Chapter 3

Dealing with Variables

In Chapter 1, the distinction between Variables and Attributes was explained. In this chapter, we shall introduce ways of expressing numerical data as values of some variable. Following that we shall address how to draw conclusions from the data.

A variable is usually expressed in terms of some continuous measurement such as feet, mmol/L [blood sugar content], etc. That type of continuous variable can take any value in a range and the number of significant figures is limited only by the accuracy available to the measuring technique involved.

Metrics

There are, on the other hand, semi-quantified metrics, which can also be treated as variables. Sometimes, we cannot measure some specific feature of interest but have to rely on subjective judgement. An example would be asking a patient to assess perceived pain level using a score of 1 to 5 (1 being minimal and 5 being extreme and unbearable). This type of metric has no units but, nevertheless, we can treat it much like a variable. Figure 3.1 shows an example.

DOI: 10.4324/9781003220138-3

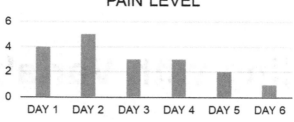

Figure 3.1 Semi-quantified metric assessing pain

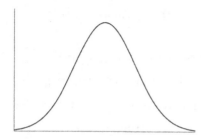

Figure 3.2 Normal (Gaussian) curve

In this chapter, we will confine ourselves to the specific distribution (spread) of a measured value which is called the NORMAL (or Gaussian) distribution. This applies where the values are equally spread on either side of the average (centre) value and form a symmetrical bell-shaped curve the shape of which is illustrated in Figure 3.2.

The two most important "measures" which are relevant to such a collection of values are

A measure of "central tendency" or "average" called the
MEAN

A measure of "scatter" or "spread" called the
STANDARD DEVIATION

The Mean

The Mean, or to be more specific, the Arithmetic Mean, is the sum of all the values divided by the total number of individual values. This is more concisely expressed as a formula:

$$\bar{X} = \frac{\Sigma Xi}{N}$$

The X with a bar on top signifies the Mean of the X values. X_i signifies each of the N values of X. The Greek letter sigma Σ signifies adding together all the X_i values.

The Standard Deviation (and Variance)

This involves taking the difference between each value of X_i and the Mean and then formulating a parameter to describe the degree of scatter. That is the standard deviation and the formula is

$$\sqrt{\frac{\Sigma(\bar{X} - Xi)^2}{N}}$$

Without the square root it is called the variance, in other words:

$$\frac{\sum(\bar{X} - Xi)^2}{N}$$

Grouped Data

Sometimes, in a set of data, we have more than one item with the same value. In order to record these more succinctly, each value is recorded once with a separate entry to indicate how many such items. The following Exercise 2 will handle the data in that form.

A Range for Each Value

Where the data is expressed as an integer, then any number in a range either side of the integer will take that integer value. Thus, in the following Exercise 2, the number 7 denotes being between 6.5 and 7.5 years of age and so on.

Making Inferences from the Mean and Standard Deviation

We have seen earlier that, where there is a symmetrical distribution, it is meaningful to state that the probability (chance) of exceeding the mean value is 50%. This much is fairly obvious. However, we are now in a position to make

Exercise 2   Obtaining the Mean and Standard Deviation

Calculate the mean and standard deviation for the following set of data. Make use of the information in Appendix 1 to do this using a spreadsheet.

Age	Number in Group
6	1
7	2
8	4
9	4
10	2
11	1

a far wider range of inferences. Figure 3.3 shows a normal curve and introduces the two ideas of **Inference** and **Confidence.**

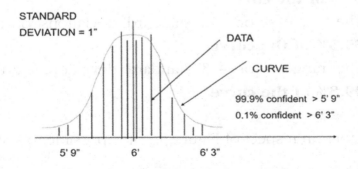

Figure 3.3   Normal curve and data

Inference means that statements based on the mathematics of that curve will hold good for the data in question, given that the data set is seen to be a good fit to the Normal distribution curve. In Chapter 8, we will look at a method of determining the "goodness of fit".

Confidence is the term used to refer to the probability (or chance) that a particular inference is true.

In the earlier statement about exceeding the mean then, for Figure 3.3, a statement might be "50% Confident that a person is more than 6 feet in height". This is borne out by the observation that 50% of the area under the normal curve lies to the right of the mean.

By a similar use of the areas under the curve, it is possible to state the likelihood (**confidence**) of being less than, greater than, or falling between, any two values. Appendix 3 provides a table of the normal distribution.

Look at Appendix 3 and, by taking differences, note that:

- The range + or − 1 standard deviation occupies **68% of the curve**
- The range + or − 2 standard deviations occupies **95.5% of the curve**
- The range + or − 3 standard deviations occupies **99.8% of the curve**

Therefore, in respect of Figure 3.3, it is possible to say that we are

■ 68% confident that a person's height is between 5' 11" and 6' 1".

■ 95.5% confident that a person's height is between 5' 10" and 6' 2".

■ 99.8% confident that a person's height is between 5' 9" and 6' 3".

Notice that these statements reflect a range and are thus referred to as double-sided inferences.

We could, however, say that one is:

■ 99.9% confident that a person's height is greater than 5' 9".

This latter statement would be referred to as a single-sided inference.

The Coefficient of Variation

We are now equipped to make use of what is known as the coefficient of variation (COV). It is simply the ratio of the standard deviation to the mean. Expressed as a percentage it provides a "feel" for the degree of scatter and hence the consistency (or otherwise) of the set of values.

A COV of 5% would mean that 68% of the values fall between + or – 5% of the mean. This would indicate a far tighter (consistent) distribution than a COV of, say, 10%.

Figure 3.4 shows graphs relating to a patient's blood sugar over a nine-day period. One shows the trend in mean sugar level and the other the trend in COV. As can

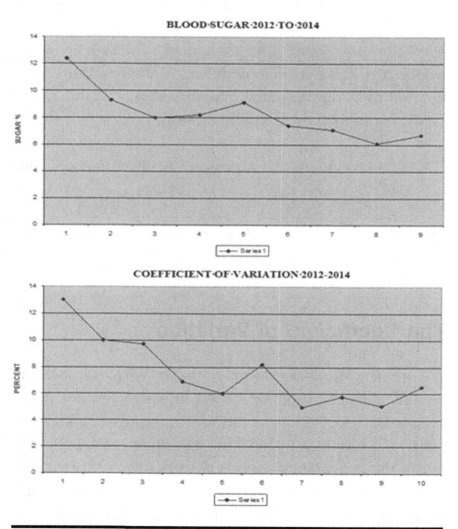

Figure 3.4   Graphs of Mean and COV blood sugar

be seen, not only has the sugar level tended to fall but so also has the COV, indicating an improvement in consistency. The latter is as important as the trend in the Mean.

Exercise 3   Mean and Coefficient of Variation

The following mmol/L blood sugar measurements were recorded by a patient. Calculate the mean and COV and comment.

13-Nov	14-Nov	15-Nov	16-Nov	17-Nov	18-Nov	19-Nov
6.90	6.87	6.83	6.90	6.33	6.63	6.67

Other Measures of Central Tendency (Types of Average)

Medians

The median is that value on either side of which 50% of the values lie. Thus,

> 1,2,3,7,8,10,11 yields a Median of **7** (note that the mean is smaller, **6**)
> This is because the distribution is skewed to the left.

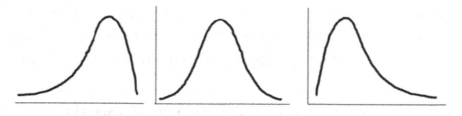

Figure 3.5 Skewed distributions from left to right: The first example, an unbiased (symmetrical) distribution, and the second example

> 1,2,4,8,10 yields a Median of **4** (note that the mean is greater, **5**)
> This is because the distribution is skewed to the right.

Figure 3.5 illustrates, from left to right:

- The first example
- An unbiased (symmetrical) distribution
- The second example

For a normal distribution:

$$\text{The Mean} = \text{The Median}$$

However, since the median provides the value which we are 50% confident to exceed, it is useful for expressing expectancy of survival.

Exercise 4 Median Life Expectancy

Following a specific diagnosis, the following table shows [to the nearest year] how long each of 50 patients survived. Calculate the median life and the likelihood of survival to each year.

No of Patients	Years of Survival
2	2
6	3
8	4
11	5
13	6
9	7
1	8

Note that each survival age signifies a range of plus or minus 0.5 year on either side of the value stated.

Geometric Mean

The geometric mean of two numbers, X and Y, is calculated as the Square Root of X times Y.

Geometric means are more meaningful than arithmetic means in two areas:

a. Geometric Growth

If the growth of some variable is geometric rather than linear (e.g., population) where the arithmetic mean would be misleading. Thus, if the population of Noddytown is 10,000 in 2010, and 20,000 in 2020, then the likely number in 2015 would be better given as:

Square Root of 10,000 × 20,000 = 14,142.
Rather than 15,000 (arithmetic mean)

In Chapter 5, this will be illustrated graphically.

b. A wide range

Assessing the most likely value when we only know a wide range (e.g., 1 to 100).

Again, the square root of 1 × 100 = 10.

This is arguably more meaningful than taking the arithmetic mean, of approximately 50, which is 50 times the smaller value but only half of the larger value.

Chapter 4

Comparing Variables

In Chapter 3, we acquired some familiarity with the normal (Gaussian) distribution and its mean and standard deviation.

The next step is to think about comparing two distributions (of the same variable) with a view to deciding if they are significantly different or whether they might both represent the same population.

Figure 4.1 illustrates the idea graphically. The two questions are

a. Are the standard deviations significantly different?

b. Are the means significantly different?

Comparing the Standard Deviations

A factor known as F (from Fisher's F distribution) is calculated as the ratio of the larger to the smaller of the two variances. Remembering that the variance is the square of the standard deviation, then:

$$F = \frac{StdDev1^2}{StdDev2^2}$$

Appendix 4 provides tables of the F distribution. In order to use them, it is necessary to make use of a number

DOI: 10.4324/9781003220138-4

(a) (b)

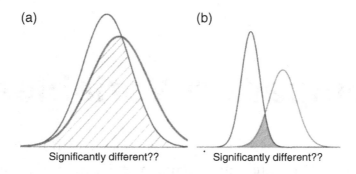

Significantly different??   Significantly different??

Figure 4.1   Comparing distributions: a) Are the standard deviations significantly different? b) Are the means significantly different?

known as the degrees of freedom. It is simply n–1, for each of the variances, where n is the number of items in each of the two distributions being considered.

Assume that there are two distributions that we wish to compare, where F = 3.5 and the two sample sizes (n) are both 10. Looking at the 2.5% table (Appendix 4b) it can be seen that the nearest entry to 9 degrees of freedom (n–1) is given by column10/row10 which gives a value of 3.72. The table tells us that there is, therefore, an approximately 2.5% chance that the two distributions have been drawn from a common population. In other words, it is 97.5% likely that they are different.

Imagine, for a moment, that the value of F were to be 5.0 then, looking at the 1% table (Appendix 4a) it can be seen that the nearest entry to 9 degrees of freedom (n–1) is 4.85. The table tells us that there is, thus, an approximately 1% chance that the distributions are from a common population. In other words, it is 99% likely that they are significantly different.

Notice that larger values of F indicate a greater significance of difference.

Comparing the Means

A factor "t" (from the Students t distribution) is calculated from the difference of the means divided by the square root of $[\text{StdDev1}^2/n_1 + \text{StdDev2}^2/n_2]$. Hence,

$$t = \frac{[\text{Mean1}-\text{Mean2}]}{\text{SQRT}[\text{StdDev1}^2/n_1 + \text{StdDev2}^2/n_2]}$$

The degrees of freedom are given by $n_1 + n_2 - 2$

Looking at the t table (Appendix 5) and, assuming that $t = 0.7$ (with 10 degrees of freedom), we can see that there is only a 50% likelihood that the distributions are significantly different. As the value of t increases, then so does the likelihood of a significant difference.

We can now proceed to the next Exercise making use of the F tables in Appendix 4 and the t Tables in Appendix 5.

Exercise 5   Comparing Two Distributions

Two treatment regimes, for Parkinson's disease, yielded the following response times in a reaction test.

For each of 6 patients before treatment:

61, 32, 37, 43, 29, 50 seconds

For each of those 6 patients after treatment:

45, 30, 32, 36, 27, 43 seconds

Determine if the variances and means are significantly different and form an opinion about the effectiveness, or otherwise, of the treatment.

Double- and Single-Sided Inferences

In Chapter 3 (Figure 3.3 and subsequent paragraphs), we saw how single-sided statements (such as "greater than", "less than") and doubled-sided statements ("between") can be made by choosing the appropriate areas of the normal distribution curve (Appendix 3). In much the same way, we can do this with the "t" distribution.

The double-sided use of "t" (which are the tables in Appendix 5) provide inferences that the difference is (or is not) significant in respect of "one mean greater" or "the other mean greater". In most cases, this is the interpretation we require. The question of whether the difference is desirable or undesirable, in the case of the treatments/options being compared, is a matter for the analyst to interpret.

Chapter 5

Presenting Data and Establishing Trends

Visual Presentation of Data

Histograms

Statistics is concerned with analysing numerical data, so the way in which the data are presented is of great importance. We are frequently told "A pictures paints a thousand words". Consider Table 5.1 which gives 100 values of a hypothetical parameter involved in some aspect of haematology. The 100 values are presented in 10 rows of 10 and in the order in which they were observed.

The appearance of the above table of values provides no picture of the way in which the values themselves are distributed. It is not easy to see which value occurs most frequently, or which value indicates the centre point of the distribution or, for that matter, whether the values are distributed symmetrically about the centre. It is very likely that we would want some idea of the degree of scatter (deviation) of the values and this "picture" is certainly not evident from the table.

DOI: 10.4324/9781003220138-5

Table 5.1   An Example of Some Data

140	155	335	211	211	198	323	294	321	235
251	271	181	227	361	306	221	252	311	165
255	255	273	293	181	259	258	236	237	211
191	294	256	273	272	220	274	256	220	301
237	244	275	258	276	220	192	192	237	241
260	221	192	292	189	301	279	281	241	231
232	220	261	281	282	242	262	225	283	264
226	284	243	265	227	208	311	245	286	266
247	246	267	289	247	227	228	268	322	248
172	335	208	209	291	269	249	249	269	192

Table 5.2 Grouping and Ranking the Data

Class Boundaries	Frequency
130–149	1
150–169	2
170–189	4
190–209	9
210–229	15
230–249	19
250–269	19
270–289	15
290–309	8
310–329	5
330–349	2
350–369	1
	100

Table 5.2, however, shows the same values arranged in a more orderly fashion. Three features distinguish the two Tables 5.1 and 5.2. In the second, table the range of values is divided into class boundaries such that all values falling within a particular class are treated as similar. Secondly, each class is indicated once only and the number of values falling within each class is indicated in the frequency column. Thirdly, the classes are ranked from lowest to highest.

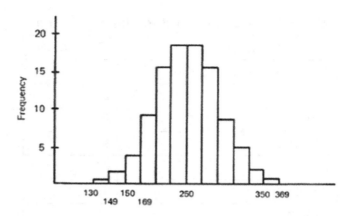

Figure 5.1   The data as a histogram

The central tendency of the values, their range and the way in which they are clustered around the central value are somewhat clearer as a result of their grouping in Table 5.2.

Better still, in Figure 5.1, this information is presented in the form of a bar chart (**HISTOGRAM**) in which each class is represented by a bar whose height is proportional to the number of values within that class.

The question arises as to how many classes should be chosen into which to allocate the values. If the number of classes is too high, relative to the number of values in the data set, then there will be too few values in each class and the histogram will appear flat as seen in Figure 5.2.

If, on the other hand, there are too few classes, then the histogram will look like Figure 5.3.

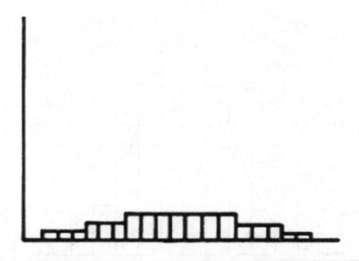

Figure 5.2   Histogram with too many classes

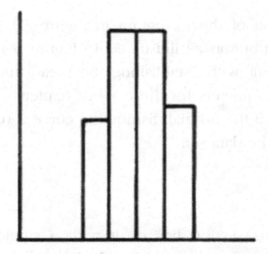

Figure 5.3   Histogram with too few classes

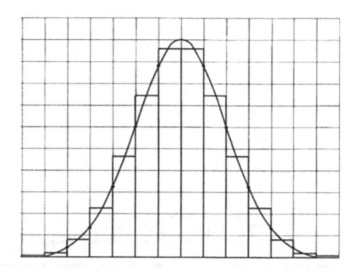

Figure 5.4   Data showing the superimposed normal distribution

The shape of the histogram in Figure 5.1 suggests a normal distribution as illustrated in Figure 5.4. Chapters 3 and 4 dealt with establishing the mean and standard deviation parameters for the curve. Chapter 8 deals with establishing if the normal distribution curve is (or is not) a good fit to the data set.

Graphs

The previous section has dealt with a single variable. Frequently we are presented with data concerning two, possibly linked, variables. The following Table 5.3 shows 24 blood pressure readings in mmHg. Again, the table alone tells us very little.

Table 5.3 24 Blood Pressure Readings

Systolic	Diastolic	Systolic	Diastolic
148	73	144	82
132	61	167	77
139	73	134	64
155	66	140	64
136	66	129	66
133	64	150	63
162	73	115	62
142	71	118	64
128	58	134	67
136	62	132	61
146	54	122	61
143	54	120	56

Figure 5.5 shows how, by representing each reading as a Y and an X value on a diagram, a visual picture of their relationship emerges.

The danger comes when we try to express the relationship more precisely by drawing a straight line (i.e., a regression line) which appears to best fit the data.

Figure 5.6 shows how several possible regression lines might each claim to serve that purpose. The technique, although useful in terms of a general picture has, nevertheless, the potential to be misleading.

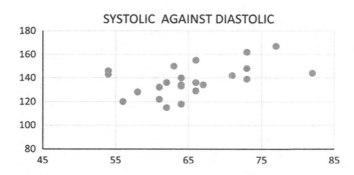

Figure 5.5   Scatter diagram

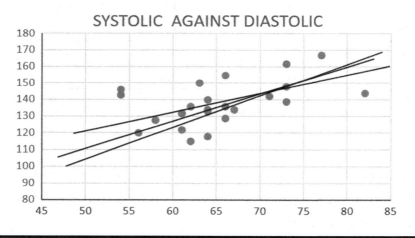

Figure 5.6   Scatter diagram with regression lines

In Chapter 8, we shall deal with correlation and regression where a more precise method of assigning the line will be explained. However, this can then to lead to the bad practice of showing the line without the scatter points, thus giving a false impression of the strength of association of the two variables.

Some Pitfalls

Supression

The following graph (Figure 5.7) shows the coefficient of variation of a patient's blood sugar over four months, giving the impression of a very significant trend of improvement. Note, however, that the vertical axis only displays a range from 12% to 21%.

Figure 5.8 shows the true picture, which is not as spectacular as might have been inferred, when supressing the zero.

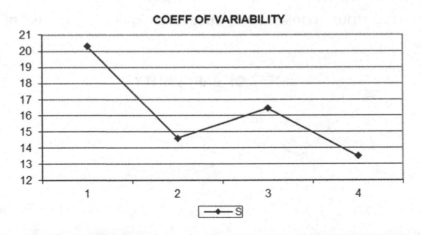

Figure 5.7   Trend in COV

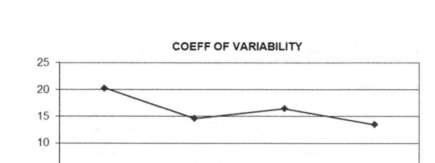

Figure 5.8 Trend with zero no longer supressed

Extrapolation

Figure 5.9 illustrates the danger of assuming that a trend will continue unaltered. The data in Figures 5.7 and 5.8 only provide four consecutive measurements. There is no

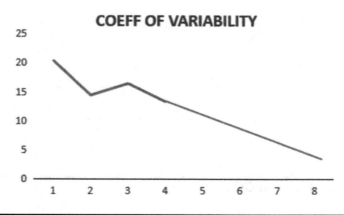

Figure 5.9 Example of extrapolation

guarantee that the trend will continue, in that the circumstances pertaining to the days beyond Day 4 may change, such as to alter the trend. Nevertheless, extrapolation of graphs is common. This practice is sometimes justified but needs to be argued (i.e., justified) for each and every case.

Logarithmic Scales

The difference between a linear and a logarithmic scale is illustrated by Figures 5.10 and 5.11. Figure 5.10 shows the growth in population of some hypothetical country. The graph illustrates, not unrealiastically, the accelerating nature of the growth.

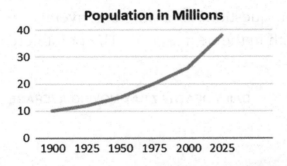

Figure 5.10   Population growth (linear scale)

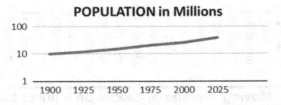

Figure 5.11   Population growth (logarithmic scale)

Figure 5.11, however, plots the same data on a logarithmic scale. That is to say equal distances, ascending the y axis, represent equal MULTIPLES rather than equal INCREMENTS. The picture is quite different. It suggests a more or less constant rate of growth.

Both representations are valid in their own way depending on whether the intention is to illustrate the absolute number (Figure 5.10) or the rate of growth (Figure 5.11).

It reminds us to be careful when looking at graphs to check what linearity applies to the axes since, improperly presented, a picture could be misleading.

Figures 5.12 and 5.13 provide a further example where, this time, the graph shows a decreasing quantity. The number in question is a moving average which will be dealt with in the next section. The point here is that the

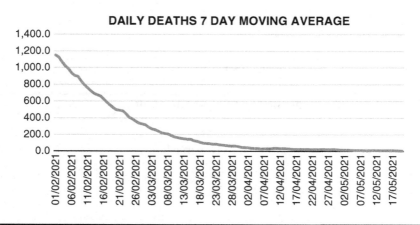

Figure 5.12   Moving average of deaths on a linear basis

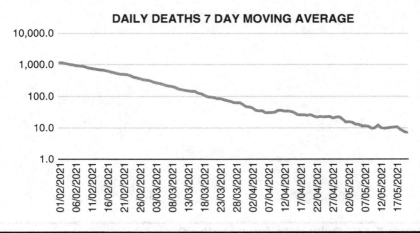

Figure 5.13   Moving average of deaths on a logarithmic basis

decrease in average UK COVID death rate, from mid-January onwards, appears to become less and less when expressed on a linear axis (Figure 5.12). The logarithmic picture (Figure 5.13), however, shows a fairly constant relative drop as time goes by.

Moving Averages

The following graph (Figure 5.14) shows the variation of daily UK fatalities during the 2020 COVID pandemic.

Daily variations (during each week) and the effect of weekends give rise to a ragged picture. If the mean (average number of fatalities) is calculated for a moving window of 7 days (Figure 5.15), the picture becomes clearer.

Figure 5.14   Daily data (unsmoothed)

Figure 5.15   Daily data (smoothed)

In Figure 5.15, a 7-day window was used due to the irregularity of the numbers published at weekends. The "weekend distortion" effect was, thus, smoothed out.

In the following example (Figure 5.16), there are 16 consecutive values of a patient's estimation of pain. The graph of raw data shows no obvious trend.

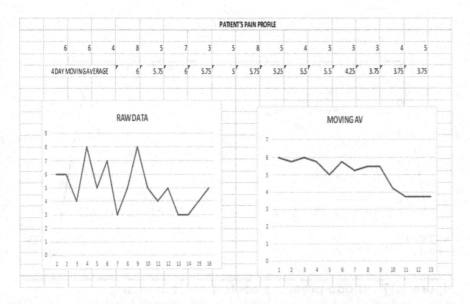

Figure 5.16 Charts of pain estimation

However, by taking a 4-day moving average a trend emerges. The first value of the moving average, in this case, is constructed by taking the average of the first 4 values (6, 6, 4, 8). The next removes the "6" value and includes the "5". The window thus moves to the right selecting 4 values at a time. The choice of a 4-day window is arbitrary and should be made having regard to the circumstances.

Control Charts

These are graphs of a chosen variable with warning and action lines to stimulate responses to undesirable variations. Figure 5.17 shows a simplified blood pressure

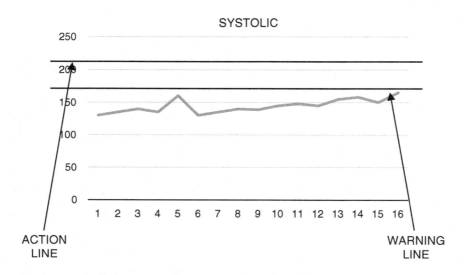

Figure 5.17 Blood pressure control chart in mmHg

chart with action and warning lines. The choice of values for each line may be a matter of judgement or, if data is available, from previous outcomes involving those levels.

Chapter 6

Dealing with Attributes

Up to now we have been dealing with Variables where the data is expressed as values of some item of interest (e.g., weight, size, blood sugar values). As we saw, at the beginning of the book, Attributes involve integers whereby we count the number of items to which the specific Attribute applies.

Binomial

Think about the following experiment with a deck of 52 playing cards. A card is removed at random, its suit noted, and then replaced. A second card is then removed, and its suit also noted. The possible outcomes of this experiment, in respect of hearts, are as follows:

H&H: H&notH; notH&H; notH&notH

where 'H' signifies a heart and 'notH' signifies a card other than a heart.

The probability of drawing a heart from a normal deck of cards is 0.25 (P). The Probability of drawing a card of some other suit is therefore:

DOI: 10.4324/9781003220138-6

$$1 - 0.25 = 0.75 \ (Q).$$

Note that $P + Q = 1$

From the multiplication rule in Chapter 2

■ The probability of drawing 2 hearts is $P \times P = P^2 = 0.0625$
■ The probability of drawing a heart as the first card and some other suit as the second card is $P \times Q = 0.1875$
■ The probability that the first card is not a heart and that the second is a heart is $Q \times P = 0.1875$
■ The probability that both are not hearts is $Q \times Q = Q^2 = 0.5625$.

Expressing this in general we can say:

■ Probability of 2 hearts P^2
■ Probability of 1 heart $2PQ$
■ Probability of 0 hearts Q^2

In the same way, for 3 cards the result will be:

■ Probability of 3 hearts P^3
■ Probability of 2 hearts $3P^2Q$
■ Probability of 1 heart $3PQ^2$
■ Probability of 0 hearts Q^3

The reader will recognise the above probabilities as the terms of the expressions $(P + Q)^2$ and $(P + Q)^3$. This leads to the general statement that if P is the probability of some random event, and if Q is $1 - P$, then the probabilities of 0, 1, 2, 3, etc., outcomes of that event, in n trials, are given by the terms of the expansion:

$$(P + Q)^n = P^n, nP^{(n-1)}Q, \frac{n(n - 1)P^{(n-2)}Q^2}{2!} \dots \dots \dots \text{ etc.}$$

This is known as the Binomial expansion. It has been included for general background (depth of knowledge) and interest. In most cases, we will be able to move on to the next section which provides a simpler approach.

Poisson

In the case of attributes (which is what we are dealing with in this Chapter) items are classified as either having the attribute or not. So for a sample of n, drawn from a population containing the P proportion, probabilities of seeing 0, 1, 2, 3, etc., are given by the terms of the above Binomial expansion. Usually, P is a very small number (e.g., 0.05 which is 5%) and n is a reasonably large number say >5. In such cases an approximation known as the POISSON DISTRIBUTION applies. The great advantage of the Poisson Distribution is that, whereas the Binomial is specified by 2 quantities (P and n), the Poisson requires only the single number given by the product of nP. We call nP the "expectation" since it is the expected average number of items, in the sample, which have the specified attribute. This does not infer that every sample will contain the quantity given by exactly n × P, indeed some will contain 0, others 1, others 2 and so on. If the expectation is given the symbol m such that m = nP then the terms of the Poisson Distribution are:

$$e^{-m}, \quad me^{-m}, \quad \frac{m^2 e^{-m}}{2!}, \quad \frac{m^n e^{-m}}{n!}$$

For each of:

| 0 | 1 | 2 | n items |

where e is a number (often used in mathematics) whose value is approximately 2.7. It is not necessary to remember the value of e, or the terms of the Poisson expansion because, as is the case with the Normal, F and t Distributions, tables are available. In the case of the Poisson Distribution these tables are often expressed in the form of a set of curves which give the probability of n or less items, having the attribute in question, plotted against m for different values of n. A little caution is necessary since some tables and curves give the probabilities of n or more occurrences whilst others give those of n or less and others the probability of exactly n. Appendix 6 gives curves of the probabilities of n or less occurrences over a range of values of m from 0.1 to 20 and for values of n from 0 to 20.

In the following example, the annual probability of fatal influenza is 1 in 4,000 which is better expressed as 2.5×10^{-4} (Chapter 9 deals with the practice of expressing numbers using exponents).

Thus, in a town of 10,000 inhabitants, the anticipated number of fatal cases is $m = 2.5$. From the curves (Appendix 6) the following probabilities can be obtained:

0 deaths	is 8%
1 or less deaths	is 29%
2 or less deaths	is 54%
3 or less deaths	is 76%

Exercise 6   Attributes Using Poisson Curves

A practice cares for 2,000 potential patients.

It is known that the incidence of stroke, in the UK, is approximately 115,000 pa (population 67 million).

What is the probability that the practice will need to treat?:

_____

0 cases?
_____
1 case?
_____
2 cases?
_____
3 cases?
_____
4 cases?
_____
5 cases?
_____

Chapter 7

Testing for Significance (Attributes)

We are frequently confronted with data relating to some attribute where measurements have been taken under two different regimes or circumstances. The following technique applies to the numbers of occurrences of an attribute and is used to compare the difference between outcomes in terms of whether that difference is significant or is merely due to chance.

It is best explained by means of the following example. Assume that we wish to test if the observed frequency of seeing each face of a die is random or whether that particular die is biased.

We calculate the value of what is known as $\chi 2$ called Chi-squared (χ is the Greek letter Chi pronounced Kai, as in the word kite). Some books refer to it as Chi- square. It is calculated by comparing the observed frequencies (O) with the frequencies which could be anticipated if the outcome were assumed to be unbiased (A). The formula is

DOI: 10.4324/9781003220138-7

$$\chi^2 = \sum \frac{(O - A)^2}{A}$$

Compare the observed values from throwing the die 60 times with what might be assumed if it were to be perfectly unbiased. The first step is to list the observed frequency of "seeing" each face. Thus, column 1 of Table 7.1 denotes the face and column 2 the number of observations. In column 3, we have entered the anticipated number of occurrences if the die were perfectly balanced, in which case the 60 throws would lead to 10 occasions per face. Column 4 records the difference between the observed (O) and the anticipated (A). Column 5 squares the difference. Column 6 divides this square by the anticipated

Table 7.1 Outcome from Sixty Throws of a Die

Face	Observed Frequency	Anticipated Frequency			
	0	A	O-A	$(O-A)^2$	$(O-A)^2/A$
1	10	10	0	0	0
2	8	10	−2	4	0.4
3	6	10	−4	16	1.6
4	11	10	1	1	0.1
5	13	10	3	9	0.9
6	12	10	2	4	0.4
					3.4

result. The sum of the column 6 values provides the value of $\chi 2$.

The $\chi 2$ value of 3.4 is compared (using the Appendix 7 table) with what might be anticipated if the result were purely random. Since there were 6 faces there are 5 independent values and this is called the "degrees of freedom" namely n-1.

Row 5 of Appendix 7 tells us that 3.4 is between 70% and 60% likelihood. Therefore, we infer that it is approximately 65% likely that the difference is due purely to chance and we might reasonably infer that the die is not biased.

The above example compared a set of outcomes with an assumed hypothesis. In Exercise 7, we will use the technique to compare to sets of outcomes resulting from different treatments

Exercise 7   Chi- Squared

Comment on the effectiveness of the following two treatment regimes for diabetes.

	TREATMENT A WITH INSULIN ONLY	TREATMENT A	
	BASE DATA	TREATMENT A	ANTICIPATED
	IMPROVEMENT	IMPROVEMENT	IF NO EFFECT
	> 10% REDUCTION	> 10% REDUCTION	TREATMENT A
	W/O TREATMENT	WITH TREATMENT	
1 MONTH	4	2	1.655172
1–2 MONTHS	15	6	6.206897
2–3 MONTHS	6	2	2.482759
>3 MONTHS	4	2	1.655172
	29	12	12

TREATMENT B WITH INSULIN & ADDITIONAL TREATMENT

	BASE DATA IMPROVEMENT > 10% REDUCTION W/O TREATMENT	TREATMENT B IMPROVEMENT > 10% REDUCTION WITH TREATMENT	ANTICIPATED IF NO EFFECT TREATMENT B
1 MONTH	4	0	2.344828
1–2 MONTHS	15	8	8.793103
2–3 MONTHS	6	5	3.517241
>3 MONTHS	4	4	2.344828
	29	17	17

Chapter 8

Correlation and Regression

Relating Two Variables

Having dealt with attributes, we now return to looking at variables and would like a technique that provides a metric for describing the strength of association between two variables.

In Figure 8.1, the scatter chart shows a number of blood pressure readings that record the systolic and diastolic blood pressures. The point of interest is whether the one is a reliable predictor of the other. The chart shows some (albeit loose) correlation between the two and the question is "what is the strength of that association". The lines drawn through the data show that the association is not exact. In fact, we could potentially draw several different lines which look equally convincing.

The technique is to fit a line to the data which best models the fit. The metric that is used to describe the "goodness of fit" to the data values is called the correlation coefficient and it is given the symbol "**r**". The formula is

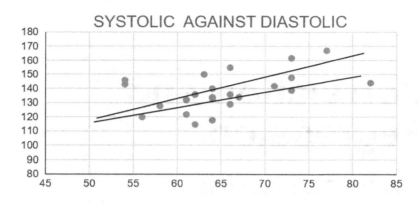

Figure 8.1   Two possible regression lines (visually)

not simple and is dealt with in References 1 and 2 (see Appendix 9).

> **r = 1** signifies 100% correlation as, for example, between water pressure and water depth
> **r = 0** signifies no correlation at all (in other words randomness of association between the two variables)

In practice, we would be looking for $r^2 > 0.95$ for a robust association (r^2 is referred to as the coefficient of determination). For those readers who would like to take the subject a little further the formula is given below:

$$\sum (x_i - x_{mean})(y_i - y_{mean})/\text{SquareRoot}[\sum (x_i - x_{mean})^2 \sum (y_i - y_{mean})^2)]$$

The Greek sigma, $\sum$, signifies sum all of the values.

Figure 8.2 is a spreadsheet set to calculate "r".

▲	A	B	C	D	E	F	G	H
1	Item	X	Y	X-Mean	Y-Mean	(X-Mean)x (Y-Mean)	(X-Mean)sq	(Y-Mean)sq
2		148	73	10.29	7.92	81.48	105.92	62.67
3		132	61	-5.71	-4.08	23.31	32.59	16.67
4		139	73	1.29	7.92	10.23	1.67	62.67
5		155	66	17.29	0.92	15.85	299.00	0.84
6		136	66	-1.71	0.92	-1.57	2.92	0.84
7		133	64	-4.71	-1.08	5.10	22.17	1.17
8		162	73	24.29	7.92	192.31	590.09	62.67
9		142	71	4.29	5.92	25.39	18.42	35.01
10		128	58	-9.71	-7.08	68.77	94.25	50.17
11		136	62	-1.71	-3.08	5.27	2.92	9.51
12		146	54	8.29	-11.08	-91.90	68.75	122.84
13		143	54	5.29	-11.08	-58.65	28.00	122.84
14		144	82	6.29	16.92	106.43	39.59	286.17
15		167	77	29.29	11.92	349.06	858.00	142.01
16		134	64	-3.71	-1.08	4.02	13.75	1.17
17		140	64	2.29	-1.08	-2.48	5.25	1.17
18		129	66	-8.71	0.92	-7.98	75.84	0.84
19		150	63	12.29	-2.08	-25.61	151.09	4.34
20		115	62	-22.71	-3.08	70.02	515.67	9.51
21		118	64	-19.71	-1.08	21.35	388.42	1.17
22		134	67	-3.71	1.92	-7.11	13.75	3.67
23		132	61	-5.71	-4.08	23.31	32.59	16.67
24		122	61	-15.71	-4.08	64.14	246.75	16.67
25		120	56	-17.71	-9.08	160.85	313.59	82.51
26								
27								
28	SUM	3305	1562			1032	3921	1114
29								
30	MEAN	137.7	65.1					
31								
32			R	0.493626				

Figure 8.2   Regression spreadsheet

As can be seen in Figure 8.2, the r value of barely 0.5 signifies a very poor correlation between the systolic and diastolic readings in respect of that particular set of data.

The regression equation is a straight-line expression of the form:

$$Y = mX + c$$

It can be constructed from the data and this is dealt with in References 1 and 2 (see Appendix 9). In addition, Appendix 1 explains the spreadsheet function for obtaining this.

Exercise 8   Correlation

Construct the spreadsheet shown in Figure 8.2 using the formulae indicated. Cell D32 (Figure 8.2) contains the equation for r provided above the figure. Enter the following data and determine the coefficient, r.

Systolic	Diastolic
147	73
156	78
160	73
189	80
173	68
157	70
139	80
157	79
165	60
156	82
175	78
158	70
161	80
180	74
145	78
162	69

(Continued)

Systolic	Diastolic
147	70
168	79
131	68
138	75
126	58
147	73
145	70
145	69

False Correlation

Beware of false correlations where a large value of r is obtained but, in fact, the two variables are only linked by virtue of some common factor.

For any two correlated events, A and B, their possible relationships include the following:

- A causes B (direct causation);
- B causes A (reverse causation);
- A and B are both caused by C;
- A causes B and B causes A;
- There is no connection between A and B; the correlation is a coincidence.

Thus, there can be no conclusion made regarding the existence or the direction of a cause-and-effect relationship only from the fact that A and B appear to be correlated. Determining whether there is an actual cause-and-effect relationship requires further investigation, even when the relationship between A and B is statistically significant.

Establishing a Normal Distribution

We may have some data giving us values for a specific variable and wish to establish if the distribution is, in fact, Normal. The following example is a list of 25 data values. The procedure is to:

- Rank the values from lowest to highest.
- Establish the median rank for each by what is known as Benard's approximation [(Rank-0.3)/(N + 0.4)].
- Use the Excel 'NORMSINV' function to express each in terms of its equivalent value as a number of standard deviations from the mean.
- Use the correlation tool (Exercise 8) to examine the goodness of fit.

The following spreadsheet shows how the NORMSINV is obtained for each of the 25 ranked items (Figure 8.3).

Rank		Median Rank	No of STDEV away from Mean "NORMSINV"	Ranked Data	
1		0.0275591	-1.917944016	1	
2		0.0669291	-1.499059227	2	
3		0.1062992	-1.246452206	2	
4		0.1456693	-1.055189681	3	
5		0.1850394	-0.896325881	3	
6		0.2244094	-0.757385572	3	
7		0.2637795	-0.631736528	4	
8		0.3031496	-0.515363243	4	
9		0.3425197	-0.405596134	4	
10		0.3818898	-0.300521332	4	
11		0.4212598	-0.198671544	5	
12		0.4606299	-0.098846884	5	
13		0.5	0	5	
14		0.5393701	0.098846884	5	
15		0.5787402	0.198671544	5	
16		0.6181102	0.300521332	6	
17		0.6574803	0.405596134	6	
18		0.6968504	0.515363243	6	
19		0.7362205	0.631736528	6	
20		0.7755906	0.757385572	7	
21		0.8149606	0.896325881	7	
22		0.8543307	1.055189681	7	
23		0.8937008	1.246452206	8	
24		0.9330709	1.499059227	8	
25		0.9724409	1.917944016	9	
			Mean	**5**	
			Std Dev	2.041241	

Figure 8.3 Establishing NORMSINV

The NORMSINV values are transferred (use paste special, values) to the following spreadsheet and correlated against the 25 data values in question. In the example, a correlation

	A	B	C	D	E	F	G	H
1	Item	X	Y	X-Mean	Y-Mean	(X-Mean)x (Y-Mean)	(X-Mean)sq	(Y-Mean)sq
2		-1.917944	1	-1.918	-4.000	7.672	3.679	16
3		-1.499059	2	-1.499	-3.000	4.497	2.247	9
4		-1.246452	2	-1.246	-3.000	3.739	1.554	9
5		-1.055190	3	-1.055	-2.000	2.110	1.113	4
6		-0.896326	3	-0.896	-2.000	1.793	0.803	4
7		-0.757386	3	-0.757	-2.000	1.515	0.574	4
8		-0.631737	4	-0.632	-1.000	0.632	0.399	1
9		-0.515363	4	-0.515	-1.000	0.515	0.266	1
10		-0.405596	4	-0.406	-1.000	0.406	0.165	1
11		-0.300521	4	-0.301	-1.000	0.301	0.090	1
12		-0.198672	5	-0.199	0.000	0.000	0.039	0
13		-0.098847	5	-0.099	0.000	0.000	0.010	0
14		0.000000	5	0.000	0.000	0.000	0.000	0
15		0.098847	5	0.099	0.000	0.000	0.010	0
16		0.198672	5	0.199	0.000	0.000	0.039	0
17		0.300521	6	0.301	1.000	0.301	0.090	1
18		0.405596	6	0.406	1.000	0.406	0.165	1
19		0.515363	6	0.515	1.000	0.515	0.266	1
20		0.631737	6	0.632	1.000	0.632	0.399	1
21		0.757386	7	0.757	2.000	1.515	0.574	4
22		0.896326	7	0.896	2.000	1.793	0.803	4
23		1.055190	7	1.055	2.000	2.110	1.113	4
24		1.246452	8	1.246	3.000	3.739	1.554	9
25		1.499059	8	1.499	3.000	4.497	2.247	9
26		1.917944	9	1.918	4.000	7.672	3.679	16
27								
28	sum	3.3E-15	125			46.4	21.9	100
29								
30	mean	0.00	5					
31								
32								
33			R	0.991143				

Figure 8.4   Establishing the regression coefficient

coefficient, r, of 99% is obtained. Thus, r^2 is 98% that indicates an excellent fit to the normal distribution. Inferences may therefore be made on the assumption that the data is normally distributed (Figure 8.4).

Chapter 9

Handling Numbers (Large and Small)

If the reader is not already familiar with manipulating numbers expressed in the "POWERS OF TEN" format, then it is important to study this chapter carefully and to practice Exercise 9.

Many of the numbers involved in this subject are either very large or very small (in other words, a number multiplied or divided by a large number, like a million).

If we use expressions such as "1 in 1,000,000" or "1 in 100,000" it becomes very cumbersome. It is, therefore, important to become familiar with the concept of "POWERS OF TEN."

Big Numbers

The "power" of ten means the number of tens multiplied together so:

10 times a year means 10 occurrences each year. There is only one "ten" so we write this as 10 to the power of 1. Maybe somewhat obvious but it sets the pattern for what follows	10^1
100 times a year is 10 × 10 each year so let us call it 10 squared or "10 to the power of 2", or even more slickly, 10 to the 2	10^2
In the same way, 1,000 each year becomes 10 x 10 x 10 or "10 to the 3" per year	10^3

In general, therefore, 10 multiplied by 10, N times is expressed as 10^N. You saw that 10 is written as 10^1. We usually omit the "1," but it has a purpose as was seen above. In the same way, the number one can be written as 10^0 (i.e., 10 to the power of zero). We sometimes need to calculate the square root in which case the exponent becomes 0.5 (i.e., 1/2). Thus, the square root of 10 is $10^{0.5}$.

Small Numbers

Now we shall look at numbers much smaller than one. This time it becomes necessary to express them in a way which tells us how many times to DIVIDE (rather than multiply) by 10.

1 in 10 years is statistically "one tenth" of an incident per year or "one divided by 10" so we call it 10^{-1}. In this context the minus signifies "divide".	10^{-1}
In the same way 1 in 100 becomes one in "10 to the 2", i.e., 10^{-2}	10^{-2}
Similarly, "1 in 1,000" becomes 10^{-3}	10^{-3}
Continuing in that way until we talk about 1 in 1,000,000 as 10^{-6}	10^{-6}

In general, therefore, 1 divided by 10, N times, is expressed as 10^{-N}.

Multiplying and Dividing

Having mastered this shorthand notation, not only do very large and very small numbers become easier to write but the arithmetic becomes simpler. In the new notation:

100 years is written as	10^2
10 such periods of time is written as	10^1
The total number of years is therefore 100 × 10	
which is, of course 1,000, or:	10^3

Notice how the upper-case numbers (known as the exponents) are added when we multiply (i.e., 2 + 1 = 3).

100 bottles is written as	10^2
10 bottles per crate is written as	10^1
The total number of crates is thus	
100 divided by 10 which is, of course, 10, or	10^1

Notice how the upper-case numbers (known as the exponents) are subtracted when we divide (i.e., 2 − 1 = 1).

Some Examples

1,000 groups of 100 people	$10^3 \times 10^2$	10^5 people
30 items for 300 years	$3\ 10^1 \times 3\ 10^2$	$9\ 10^3$ item-years
500 bottles divided into 20 crates	$5\ 10^2\ /\ 2\ 10^1$	$2.5\ 10^1$ which is 25 bottles
300 deaths from scalding in a population of 60,000,000	$3\ 10^2\ /\ 6\ 10^7$	**$0.5\ 10^{-5}$** which is $5\ 10^{-6}$ per head of population

Notice, in the last example, how we made the **0.5** bigger (by ten) and the **10^{-5}** smaller by (ten) in order to make the first part of the number a sensible digit. In the same way:

$500\ 10^4$ is	$5\ 10^6$
$0.25\ 10^6$ is	$2.5\ 10^5$
$300\ 10^{-6}$ is	$3\ 10^{-4}$
$0.3\ 10^{-3}$ is	$3\ 10^{-4}$

Using the above rules makes handling large and small numbers, and their arithmetic, much simpler. In spreadsheets the 10^{-3} can be entered as 0.001 but, also, as 1E-3. In calculators, it is usually 1[EXP]-3, the [EXP] being one of the keys.

Familiarity with Negative Exponents

Remember:

$3\ 10^{-4}$ is SMALLER (not BIGGER) than $3\ 10^{-3}$.

Because the minus denotes the number of tens "on the bottom line" and therefore the larger the number in the exponent the smaller the result.

$3 \ 10^{-4}$ DIVIDED by 10 is $3 \ 10^{-5}$ – in other words the number in the negative exponent gets bigger to make the resulting number smaller.

Exercise 9a   Large and Small Numbers

Complete the following table:

	Write the Answer as a Number Followed by 10^x
500 groups of 50 people	? people
50 men for 200 years	? man-years
500 bottles divided into 25 crates	? per crate
200 deaths from scalding in a population of 50,000,000	? chance of scalding
3600 patients awaiting 12 beds	? patients per bed
1500 influenza fatalities in a population of 5,000,000	? chance of becoming a fatality
1 in 500 suffer from diabetes in a population of 75,000	? patients

Exercise 9b   Large and Small Numbers

Now try the following, without the aid of a calculator. It is a valuable skill to be able manipulate such numbers mentally.

Complete the following table:

	Write the Answer as a Number Followed by 10^x
$10^6 / 5$	?
$6 \; 10^4 / 3 \; 10^2$	?
$5 \; 10^{-3} / 2.5 \; 10^{-1}$	?
$9 \; 10^{-1} / 3 \; 10^{-2}$	?
$2 \; 10^{-3} / 4 \; 10^{-2}$	?
$10^{-4} / 5 \; 10^{-5}$	?
$3 \; 10^9 / 6 \; 10^7$	?

Chapter 10

An Introduction to Risk

The purpose of this chapter is to provide a wider perspective by introducing the risk of fatality from scenarios other than those arising from medical issues. It provides a comparison between the background occupational and leisure related risks, to which we are exposed, and those relating to health. The following rates and probabilities are, by their nature, approximate estimates. References 3 and 4 (Appendix 9) also deal with this aspect.

Individual Risk of Fatality (per Exposure to an Activity)

Table 10.1 (UK figures) shows a number of risks on the basis of the fatality rate per single exposure to an activity.

DOI: 10.4324/9781003220138-10

Table 10.1 Fatality Risks per Single Exposure to an Activity

Flu Vaccination (over 65 years old) *(based on 70% coverage of approx 12 10⁶ over 65s with zero fatalities at 90% confidence*)*	$<3\ 10^{-7}$
Commercial aviation – one trip	$1\ 10^{-7}$
Scuba diving (one dive)	$5\ 10^{-6}$
Parachute/skydiving jump	$6\ 10^{-6}$
One-hour helicopter ride (private flying)	$7\ 10^{-6}$
General anaesthetic	$1\ 10^{-5}$
Childbirth	$8\ 10^{-5}$
Gastrointestinal endoscopy	$1\ 10^{-4}$

Notes

* Notice the inference of a probability (or less), stated at some confidence level. This makes further use of the Chi-squared technique. The "probability or less" is calculated as $\chi 2\ /2T$ where T is the aggregate number of exposures to the risk and $\chi 2$ is obtained from $n = 2(k+1)$ degrees of freedom (for k occurrences) and at a probability of (1-confidence). In the above example, $\chi 2$ is found in Appendix 7, from $n = 2(0+1) = 2$ and probability 0.1, namely 4.61. Thus, $4.61/(2 \times 70\% \times 12\ 10^6) = <3\ 10^{-7}$. The same formula may be used to infer a rate, in which case T becomes the aggregate time of exposure. The technique is fully explained in Reference 4 (Appendix 9).

Note, also, that an inference based on zero failures merely establishes a maximum probability or rate (i.e., ceiling) below which the actual value is likely to fall. It might well be even lower, but the inference is limited by the size of the sample available.

Individual Risk of Fatality (per Annum)

Table 10.2 (UK figures) shows a number of risks on the basis of the fatality rate per annum.

Table 10.2   Fatality Risks per Annum

Major accidents (e.g., Flixborough, Kegworth)	$1\ 10^{-6}$
Carbon monoxide poisoning	$1\ 10^{-6}$
Working in an office	$1\ 10^{-6}$
Food poisoning	$3\ 10^{-6}$
Homicide	$1\ 10^{-5}$
Working in Industry (general)	$1.5\ 10^{-5}$
Poisoning general (excluding drugs)	$2\ 10^{-5}$
Road transport (car)	$5\ 10^{-5}$
Asbestos related cancer (mesothelioma)	$6\ 10^{-5}$
Suicide	$1\ 10^{-4}$
All random accidents (non-medical) *(e.g., occupational, sport, transport, homicide, etc.)*	$2\ 10^{-4}$
Influenza/pneumonia (non-COVID)	$2\ 10^{-4}$
Covid (UK) 2020 (50% might have died from other causes)	$5\ 10^{-4}$
Cancer	$2\ 10^{-3}$

Fatal Accident Frequency (FAFR)

Comparing different activities on the basis of fatalities "per person per annum" is not always realistic because the "participating time" is not the same for all activities.

It is a credible approximation to say that we are all exposed to the risks in Table 10.2 for equal amounts of time. However, to compare motor transport and private flying in that way would be misleading when we spend say 500 hours per annum in the car and maybe 10 hours per annum flying. The traditional way of comparing risks for that type of activity is by what is known as the FAFR. That is the number of fatalities per 100 million hours of exposure to that activity.

Table 10.3, compiled from a number of sources, compares some activities and pastimes.

Maximum Tolerable Risk

When assessing risk in industrial settings the concept of Maximum Tolerable Risk is used. The term "Maximum Tolerable" implies that under no circumstances would that activity be allowed to continue if the target is not met. Any higher risk is labelled Intolerable. It is then necessary to demonstrate whether the target is met and, if so, by a sufficient margin to render the cost of further risk reduction not justified. The same principle applies in medicine.

Table 10.3 FAFRs (Fatalities per 10^8 Hours)

Activity	FAFR
Miscellaneous:	
Staying at home	3
Struck by lightning	0.04
Transport:	
Motorcycle	450
Rail	3
Bus	1
Pedestrian (walking)	10
Motor car	12
Canoeing	400
Commercial air (worldwide)	3
Recreation:	
Horse riding	500
Rock climbing	600
Private helicopter flying	630
Private fixed wing flying	750
Gliding	670
Scuba diving	670
Swimming	1,000

The justification for expenditure on further risk reduction is dealt with in the next section.

Typical values used in industry are shown in Table 10.4.

Table 10.4   Typical Maximum Tolerable Risk Targets

Maximum Tolerable Risk for a single fatality scenario (where exposure is voluntary (i.e., employee)	10^{-4} pa
Maximum Tolerable Risk for a multi-fatality scenario (here exposure is voluntary (i.e., employee)	10^{-5} pa
Maximum Tolerable Risk for a single fatality scenario (where exposure is involuntary (i.e., public)	10^{-5} pa
Maximum Tolerable Risk for a multi-fatality scenario (where exposure is involuntary (i.e., public)	10^{-6} pa

Variations to the above targets apply to multi-risk scenarios and to different numbers of fatalities. They are dealt with in Reference 4 (See Appendix 9).

As Low as Reasonably Practicable (ALARP)

This concept imposes the obligation to continue reducing the risk below the Maximum Tolerable level depending on the cost of the proposed risk reduction. In order to apply the concept, it becomes necessary to agree a target Cost per Life Saved. A typical value used in respect of industrial and many other hazards is £2,000,000.

The calculation involves assessing the cost of a proposed risk reduction measure and then calculating, based on the predicted number of fatalities avoided, the cost for each life saved. Where this exceeds £2,000,000 the expenditure is deemed not be justified and where it is below £2,000,000 then that expenditure is called for.

The concept of a cost per life saved (often referred to as cost per fatality avoided) of £2,000,000 infers that a saved life, having avoided the hazard in question, is granted the remainder of his/her remaining life span (say an average of 35 years). However, for the purposes of evaluating medical expenditure, the life saved may lead to an additional span of more restricted duration. If, for the sake of argument, a year of extended life is obtained, then the equivalent threshold becomes £2,000,000/35 = £57,000. It so happens that a value of £60,000 per statistical year of life (SLY) is, in fact, used in many health-related justifications. This SLY needs, therefore, to be calculated as the cost of the procedure divided by the anticipated number of extended years of life.

A medical example might be the cost of a cervical smear test which is in the order of £100. It is estimated that this procedure saves 5,000 lives per year and leads to 5 additional years of life. Approximately 3,000,000 tests are conducted annually. Based on those estimates the SLY is

$$3 \ 10^6 \times £100/[5\,000 \times 5] = £12\,000$$

Based on the above argument, against a criterion of £60,000, the cost of the risk reduction is well justified. The reader may care to research the numbers used in the above example and form his/her own opinion.

Exercise 10   SLY/Cost Per Life Saved

The following approximate figures are mooted for bowel cancer screening. The cost of a home testing kit, and processing, is approximately £150. 50% of kits supplied receive a response. Ten percent of responses lead to a diagnosis of advanced adenoma. Sixty-seven percent of those diagnosed survive for at last 2 years. Comment on the cost effectiveness of the screening programme.

Variable versus Constant Rates

In this chapter, we have dealt with rates and probabilities of fatality. It is worth noting that the concept of a rate parameter, to describe the distribution of events, implies a constant rate (in other words random events). Were this not to be the case, then the concept of rate becomes meaningless just as does expressing the speed of a journey between London and Aberdeen (say 600 miles in 10 hours) as 60 mph. In reality the speed would have been continuously varying and the 60 mph parameter is only a metric for describing the trip. To deal with cases where the rate is not constant, it is necessary to make use of a more complex model, known as the Weibull distribution. This will involve more than one parameter one of which, known as the shape parameter, indicates whether the rate of occurrence is increasing, constant or decreasing. Weibull analysis is slightly beyond the scope of this book but is dealt with fully in Reference 4 (Appendix 9) and, also, in the COMPARE software package available from the author.

Chapter 11

A Final Word

I have tried to show, in this book, some fairly simple techniques for drawing conclusions from quantified physical measurements. However, let us beware of the short-sightedness which leads to a blinkered treatment of numerical data. There is a danger in running away with the conclusions as if they were the only factors which impinge on the situation in question. In medical applications, this would be the patient's overall health. I cannot stress, too strongly, the need for a holistic approach to health care. The role of the specialist is paramount but, nevertheless, it is equally important to take a wide view of the parameters involved and to seek a balance between different treatment options.

I believe the same might be said of all scientific and engineering disciplines.

If you have worked through the book, and attempted the exercises, I hope you will have gained a fair grasp of the basic principles of statistical sampling and inference. I have attempted to explain the techniques with little more than simple arithmetic. They can all be applied using the tables and curves provided.

DOI: 10.4324/9781003220138-11

If I have removed some of the mystery and suspicion which surrounds this subject and if you feel better equipped to challenge numerical claims and statements and are able to apply some of the techniques, then our mutual efforts have not been in vain.

Appendices

Appendix 1: Manipulating Numbers in Spreadsheets

It is useful to be familiar with the following manipulations when using spreadsheets for statistical calculations. These instructions relate to the spreadsheets on the following pages. The precise keystrokes may differ slightly according to the spreadsheet version being used. However, the following will point the reader in the right direction.

Arithmetic Functions

Cells B1 to B6 contain 6 items of data, namely the values 2,4,6,4,6,2.

Cell B8 Adds them [=SUM(then block the 6 cells and press return]

Cell B9 returns the average [=AVERAGE(then block the 6 cells and press return]

Cell B 10 Multiplies the 6 values together [=PRODUCT(then block the 6 cells and press return]

Cell B11 returns the square of cell B9, namely [B9^2]

Cell B12 returns the square root of cell B11, namely [B9^0.5]

[Note that a square root is the number raised to the power of the reciprocal. Thus ½ =0.5]

Cell B13 returns the standard deviation of the 6 values, namely,

[= STDEV(then block the 6 cells and press return]

Copy–Paste

Cells D8–D12 were generated by blocking cells B8–B12, Control C [to copy] and then click D8, Control V [to paste]. Note that the results for D8 and D10 are zero. This is because EXCEL will have changed the B column designation to D and we are now manipulating the data in D1–D6 which is, of course, zero. Cells D, D11, D12 and D13 return #DIV/0! Because they involve a divide by cell, in the D column, which contains 0.

Copy–Paste Special

Colum E, however, is generated by a similar process but NOT using Control V [to paste], Instead, following the Control C operation, block cell E8 and select paste special and select "values". Thus, the actual values, rather than the amended formulae, are brought across.

	A	B	C	D	E	F
1		2				
2		4				
3		6				
4		4				
5		6				
6		2				
7				PASTE !	PASTE VALUES	
8	SUM	24		0	24	
9	AVERAGE	4		#DIV/0!	4	
10	PRODUCT	2304		0	2304	
11	SQUARE	16		#DIV/0!	16	
12	SQ ROOT	4		#DIV/0!	4	
13	STD DEV	1.78885		#DIV/0!		

Spreadsheet for the Chapter 2 example on Conditional Probabilities

A useful adjunct to Chapter 2 is the use of a spreadsheet to calculate the effectiveness of both positive and negative results of tests for some condition (as for example cancer). In cell B1, the probability of suffering from the condition in question is entered as a percentage. This may well have been estimated as the proportion of the population so effected. Cells D4, D5, E4, E5 will be set according to the known results pertaining to true and false positives and negatives. The following reflects the equations given in Chapter 2

Cell D8 is D4 * B1.

Cell D9 is E4 * (1 – B1).

Cell D10 is D8 + D9.

Cell F12 is D8/D10

Cell D15 is $E5*(1-B1)$,

Cell D16 is $D5*B1$,

Cell D17 is $D15+D16$.

Cell F19 is $D15/D17$.

PROBABILITY	1.00%			
OF CANCER			HAS	HAS NO
			CANCER	CANCER
		TEST POS	90%	10%
		TEST NEG	10%	90%
		TRUE POS	0.9%	
		FALSE POS	9.9%	
		ANY POSITIVE	10.8%	
		Prob'y True/Prob'y Any Positive		8.3%
		TRUE NEG	89.1%	
		FALSE NEG	0.1%	
		ANY NEGATIVE	89.2%	
		Prob'y True/Prob'y Any Negative		99.9%

Regression/Correlation

A useful adjunct to Chapter 8 is the use of a spreadsheet to generate a regression line for relating a set of two variables. It also provides an estimate of the correlation coefficient ("r") based on a least-squares calculation.

The following illustration shows a spreadsheet with two columns of data (X and Y). We proceed as follows:

Block the two columns A and B.

Insert, Chart, Scatter

Right click on any one dot and select add trend line

Right click on the line and select r^2 and select display equation

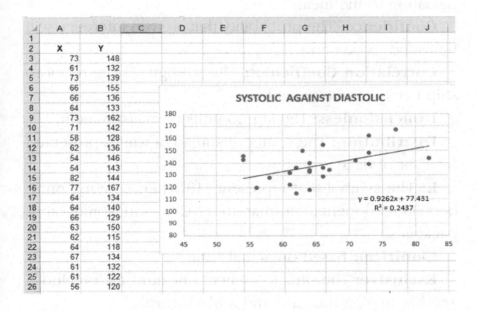

As can be seen from the above example, the correlation is very poor. In fact, it suggests that there is no linking factor between the two variables. Note also that the $Y = mX + C$ straight line equation, describing the line, is provided.

Appendix 2: Brief Glossary of Terms

À priori: Knowing in advance, rather than from observation.

Attribute: A state that either applies or does not (e.g., a red card or not)

Coefficient of Variation: The ratio of the standard deviation to the mean.

Confidence: The probability that a given inference will be found to be correct.

Correlation Coefficient: The strength of the relationship between two variables.

Dimensionless: Having no units of measurement.

Distribution: The scatter or spread of values presented by a set of data.

E: The symbol that represents 10 when entering numbers in exponential format to EXCEL and some other packages.

Empirical: Based on actual data.

Exclusive: The situation where one attribute precludes another (e.g., a red card and a black card).

EXP: The symbol that represents 10 when entering numbers in exponential format into a calculator.

FAFR: Fatal Accident Frequency (usually expressed per 100 million hours of exposure/participation in an activity).

Independence: The situation where the probability of one specified event has no influence on the probability of another specified event.

Inference: The act of predicting the probability of some outcome based on knowledge of the data in question.

Mean (arithmetic): Average value based on the sum of the values of the items divided by the number of items.

Mean (geometric): Average value based on the product of the values of the items to the power of the reciprocal of the number of items.

Median: The value, either side of which, 50% of the items, in a distribution, are to be found.

Metric: A numerical measure.

Parameter: Any feature of interest which is measurable (e.g., height, weight, blood pressure).

Population: The total number of items from which a sample is drawn.

Probability: The proportion of items which represent the factor in question (in other words, the ratio of the number of occurrences to the number of trials). *Note: This might involve a sample or a complete population.*

Random: A selection in which each item of a set has an equal probability of being chosen.

Regression: The process of establishing the mathematical relationship between two variables.

Risk: Probability or rate of some hazardous event.

Skewed Distribution A non-symmetrical distribution of values which tails (left or right).

Standard Deviation: A measure of spread given by the square root of the sum of the squares of the differences between each individual item and the mean.

Units: The values by which variables are described/measured (e.g., feet, pounds)

Variable: Any measurable feature which may take a range of values.

Variance: The square of the Standard Deviation.

Appendix 3: Normal Distribution Table

x_i	Probability	x_i	Probability
$\overline{x}$ - 3.0 σ	0.999	$\overline{x}$ + 0 σ	0.500
$\overline{x}$ - 2.9 σ	0.998	$\overline{x}$ + 0.1σ	0.460
$\overline{x}$ - 2.8 σ	0.997	$\overline{x}$ + 0.2 σ	0.421
$\overline{x}$ - 2.7 σ	0.997	$\overline{x}$ + 0.3 σ	0.382
$\overline{x}$ - 2.6 σ	0.995	$\overline{x}$ + 0.4 σ	0.345
$\overline{x}$ - 2.5 σ	0.994	$\overline{x}$ + 0.5 σ	0.308
$\overline{x}$ - 2.4 σ	0.992	$\overline{x}$ + 0.6 σ	0.274
$\overline{x}$ - 2.3 σ	0.989	$\overline{x}$ + 0.7 σ	0.242
$\overline{x}$ - 2.2 σ	0.986	$\overline{x}$ + 0.8 σ	0.212
$\overline{x}$ - 2.1 σ	0.982	$\overline{x}$ + 0.9 σ	0.184
$\overline{x}$ - 2.0 σ	0.977	$\overline{x}$ + 1.0 σ	0.159
$\overline{x}$ - 1.9 σ	0.971	$\overline{x}$ + 1.1 σ	0.136
$\overline{x}$ - 1.8 σ	0.964	$\overline{x}$ + 1.2 σ	0.115
$\overline{x}$ - 1.7 σ	0.955	$\overline{x}$ + 1.3 σ	0.097
$\overline{x}$ - 1.6 σ	0.945	$\overline{x}$ + 1.4 σ	0.081
$\overline{x}$ - 1.5 σ	0.933	$\overline{x}$ + 1.5 σ	0.067
$\overline{x}$ - 1.4 σ	0.919	$\overline{x}$ + 1.6 σ	0.055
$\overline{x}$ - 1.3 σ	0.903	$\overline{x}$ + 1.7 σ	0.045
$\overline{x}$ - 1.2 σ	0.885	$\overline{x}$ + 1.8 σ	0.036
$\overline{x}$ - 1.1 σ	0.864	$\overline{x}$ + 1.9 σ	0.029
$\overline{x}$ - 1.0 σ	0.841	$\overline{x}$ + 2.0 σ	0.023
$\overline{x}$ - 0.9 σ	0.816	$\overline{x}$ + 2.1 σ	0.018
$\overline{x}$ - 0.8 σ	0.788	$\overline{x}$ + 2.2 σ	0.014
$\overline{x}$ - 0.7 σ	0.758	$\overline{x}$ + 2.3 σ	0.011
$\overline{x}$ - 0.6 σ	0.726	$\overline{x}$ + 2.4 σ	0.008
$\overline{x}$ - 0.5 σ	0.692	$\overline{x}$ + 2.5 σ	0.006
$\overline{x}$ - 0.4 σ	0.655	$\overline{x}$ + 2.6 σ	0.005
$\overline{x}$ - 0.3 σ	0.618	$\overline{x}$ + 2.7 σ	0.003
$\overline{x}$ - 0.2 σ	0.579	$\overline{x}$ + 2.8 σ	0.003
$\overline{x}$ - 0.1 σ	0.540	$\overline{x}$ + 2.9 σ	0.002
$\overline{x}$ - 0 σ	0.500	$\overline{x}$ + 3.0 σ	0.001

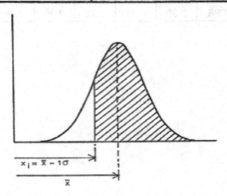

$$x_i = \overline{x} - 1\sigma$$
$$\overline{x}$$

Appendix 4a: 0.5% & 1% Points of the F Distribution

These F tables are "two-sided" in that they assess the sig nificance of there being a difference between the variances, bu not distinguishing on which side they differ.

		0.5% Probability								
		Degrees of freedom (larger std. dev.)								
		1	2	3	4	5	6	10	24	∞
Degrees of freedom (smaller std. dev.)	1	16200	20000	21600	22500	23100	23400	24200	24900	25500
	2	199	199	199	199	199	199	199	199	199
	3	55.6	49.8	47.5	46.2	45.4	44.8	43.7	42.6	41.8
	4	31.3	26.3	24.3	23.2	22.5	22.0	21.0	20.0	19.3
	5	22.8	18.3	16.5	15.6	14.9	14.5	13.6	12.8	12.1
	6	18.6	14.5	12.9	12.0	11.5	11.1	10.3	9.47	8.88
	10	12.8	9.43	8.08	7.34	6.87	6.54	5.85	5.17	4.64
	20	9.94	6.99	5.82	5.17	4.76	4.47	3.85	3.22	2.69
	30	9.18	6.35	5.24	4.62	4.23	3.95	3.34	2.73	2.18
	40	8.83	6.07	4.98	4.37	3.99	3.71	3.12	2.50	1.93
	∞	7.88	5.30	4.28	3.72	3.35	3.09	2.52	1.90	1.00

		1% Probability								
		Degrees of freedom (larger std. dev.)								
		1	2	3	4	5	6	10	24	∞
Degrees of freedom (smaller std. dev.)	1	4050	5000	5400	5620	5760	5860	6060	6230	6370
	2	98.5	99.0	99.2	99.2	99.3	99.3	99.4	99.5	99.5
	3	34.1	30.8	29.5	28.7	28.2	27.9	27.2	26.6	26.1
	4	21.2	18.0	16.7	16.0	15.5	15.2	14.5	13.9	13.5
	5	16.3	13.3	12.1	11.4	11.0	10.7	10.1	9.47	9.02
	6	13.7	10.9	9.78	9.15	8.75	8.47	7.87	7.31	6.88
	10	10.0	7.56	6.55	5.99	5.64	5.39	4.85	4.33	3.91
	20	8.10	5.85	4.94	4.43	4.10	3.87	3.37	2.86	2.42
	30	7.56	5.39	4.51	4.02	3.70	3.47	2.98	2.47	2.01
	40	7.31	5.18	4.31	3.83	3.51	3.29	2.80	2.29	1.80
	∞	6.63	4.61	3.78	3.32	3.02	2.80	2.32	1.79	1.00

Appendix 4b: 2.5% & 5% Points of the F Distribution

These F tables are "two-sided" in that they assess the significance of there being a difference between the variances, but not distinguishing on which side they differ.

		2.5% Probability								
		Degrees of freedom (larger std. dev.)								
		1	2	3	4	5	6	10	24	∞
Degrees of freedom (smaller std. dev.)	1	648	799	864	900	922	937	969	997	1020
	2	38.5	39.0	39.2	39.2	39.3	39.3	39.4	39.5	39.5
	3	17.4	16.0	15.4	15.1	14.9	14.7	14.4	14.1	13.9
	4	12.2	10.6	9.98	9.60	9.36	9.20	8.84	8.51	8.26
	5	10.0	8.43	7.76	7.39	7.15	6.98	6.62	6.28	6.02
	6	8.81	7.26	6.60	6.23	5.99	5.82	5.46	5.12	4.85
	10	6.94	5.46	4.83	4.47	4.24	4.07	3.72	3.37	3.08
	20	5.87	4.46	3.86	3.51	3.29	3.13	2.77	2.41	2.09
	30	5.57	4.18	3.59	3.25	3.03	2.87	2.51	2.14	1.79
	40	5.42	4.05	3.46	3.13	2.90	2.74	2.39	2.01	1.64
	∞	5.02	3.69	3.12	2.79	2.57	2.41	2.05	1.64	1.00

		5% Probability								
		Degrees of freedom (larger std. dev.)								
		1	2	3	4	5	6	10	24	∞
Degrees of freedom (smaller std. dev.)	1	161	199	216	225	230	234	242	249	254
	2	18.5	19.0	19.2	19.2	19.3	19.3	19.4	19.5	19.5
	3	10.1	9.55	9.28	9.12	9.01	8.94	8.79	8.64	8.53
	4	7.71	6.94	6.59	6.39	6.26	6.16	5.96	5.77	5.63
	5	6.61	5.79	5.41	5.19	5.05	4.95	4.74	4.53	4.36
	6	5.99	5.14	4.76	4.53	4.39	4.28	4.06	3.84	3.67
	10	4.96	4.10	3.71	3.48	3.33	3.22	2.98	2.74	2.54
	20	4.35	3.49	3.10	2.87	2.71	2.60	2.35	2.08	1.84
	30	4.17	3.32	2.92	2.69	2.53	2.42	2.16	1.89	1.62
	40	4.08	3.23	2.84	2.61	2.45	2.34	2.08	1.79	1.51
	∞	3.84	3.00	2.60	2.37	2.21	2.10	1.83	1.52	1.00

Appendix 4c: 10% & 25% Points of the F Distribution

These F tables are "two-sided" in that they assess the significance of there being a difference between the variances, but not distinguishing on which side they differ.

		10% Probability								
		Degrees of freedom larger (std. dev.)								
		1	2	3	4	5	6	10	24	∞
Degrees of freedom (smaller std. dev.)	1	39.9	49.5	53.6	55.8	57.2	58.2	60.2	62.0	63.3
	2	8.53	9.00	9.16	9.24	9.29	9.33	9.39	9.45	9.49
	3	5.54	5.46	5.39	5.34	5.31	5.28	5.23	5.18	5.13
	4	4.54	4.32	4.19	4.11	4.05	4.01	3.92	3.83	3.76
	5	4.06	3.78	3.62	3.52	3.45	3.40	3.30	3.19	3.10
	6	3.78	3.46	3.29	3.18	3.11	3.05	2.94	2.82	2.72
	10	3.29	2.92	2.73	2.61	2.52	2.46	2.32	2.18	2.06
	20	2.97	2.59	2.38	2.25	2.16	2.09	1.94	1.77	1.61
	30	2.88	2.49	2.28	2.14	2.05	1.98	1.82	1.64	1.46
	40	2.84	2.44	2.23	2.09	2.00	1.93	1.76	1.57	1.38
	∞	2.71	2.30	2.08	1.94	1.85	1.77	1.60	1.38	1.00

		25% Probability								
		Degrees of freedom (larger std. dev.)								
		1	2	3	4	5	6	10	24	∞
Degrees of freedom (smaller std. dev.)	1	5.83	7.50	8.20	8.58	8.82	8.98	9.32	9.63	9.85
	2	2.57	3.00	3.15	3.23	3.28	3.31	3.38	3.43	3.48
	3	2.02	2.28	2.36	2.39	2.41	2.42	2.44	2.46	2.47
	4	1.81	2.00	2.05	2.06	2.07	2.08	2.08	2.08	2.08
	5	1.69	1.85	1.88	1.89	1.89	1.89	1.89	1.88	1.87
	6	1.62	1.76	1.78	1.79	1.79	1.78	1.77	1.75	1.74
	10	1.49	1.60	1.60	1.59	1.59	1.58	1.55	1.52	1.48
	20	1.40	1.49	1.48	1.47	1.45	1.44	1.40	1.35	1.29
	30	1.38	1.45	1.44	1.42	1.41	1.39	1.35	1.29	1.23
	40	1.36	1.44	1.42	1.40	1.39	1.37	1.33	1.26	1.19
	∞	1.32	1.39	1.37	1.35	1.33	1.31	1.25	1.18	1.00

Appendix 5: The t Distribution

These "t" tables are "single-sided" in that they assess the significance of one mean being different in a specified direction.

α is the likelihood that the value occurs by chance if the means are indeed the same. Thus $(1 - \alpha)$ is the probability that they are different.

				Probability of a larger value					
α	**50%**	**40%**	**20%**	**10%**	**5%**	**2.50%**	**1%**	**0.50%**	**0.10%**
1	1.000	1.376	3.078	6.131	12.710	25.452	63.657	>>	>>
2	0.816	1.061	1.886	2.920	4.303	6.205	9.925	14.089	31.598
3	0.765	0.978	1.638	2.353	3.182	4.176	5.841	7.453	12.941
4	0.741	0.941	1.533	2.132	2.776	3.495	4.604	5.598	8.610
5	0.727	0.920	1.476	2.015	2.571	3.163	4.032	4.773	6.859
6	0.718	0.906	1.440	1.943	2.447	2.969	3.707	4.317	5.959
7	0.711	0.896	1.415	1.895	2.365	2.841	3.499	4.029	5.405
8	0.706	0.889	1.397	1.860	2.306	2.752	3.355	3.832	5.041
9	0.703	0.883	1.383	1.833	2.262	2.685	3.250	3.690	4.871
10	0.770	0.879	1.372	1.812	2.228	2.634	3.169	3.581	4.587
11	0.697	0.876	1.363	1.796	2.201	2.593	3.106	3.497	4.437
12	0.695	0.873	1.356	1.782	2.197	2.560	3.055	3.428	4.318
13	0.694	0.870	1.350	1.771	2.160	2.533	3.012	3.372	4.221
14	0.692	0.878	1.354	1.761	2.145	2.510	2.977	3.326	4.140
15	0.691	0.866	1.341	1.753	2.131	2.490	2.847	3.286	4.073
16	0.690	0.865	1.337	1.746	2.120	2.473	2.921	3.252	4.015
17	0.689	0.863	1.333	1.740	2.110	2.458	2.898	3.222	3.965
18	0.688	0.862	1.330	1.734	2.101	2.445	2.878	3.197	3.922
19	0.688	0.861	1.328	1.729	2.093	2.433	2.861	3.174	3.883
20	0.687	0.860	1.325	1.275	2.086	2.423	2.845	3.153	3.550
21	0.686	0.859	1.323	1.721	2.080	2.414	2.831	3.135	3.819
22	0.686	0.858	1.321	1.717	2.074	2.406	2.819	3.119	3.792
23	0.685	0.858	1.319	1.714	2.069	2.398	2.807	3.104	3.762
24	0.685	0.857	1.318	1.711	2.064	2.391	2.797	3.090	3.745
25	0.684	0.856	1.316	1.708	2.060	2.385	2.787	3.078	3.725
26	0.684	0.856	1.315	1.706	2.056	2.379	2.779	3.067	3.707
27	0.684	0.855	1.314	1.703	2.052	2.373	2.771	3.056	3.690
28	0.683	0.855	1.313	1.701	2.048	2.368	2.763	3.047	3.674
29	0.683	0.854	1.311	1.699	2.045	2.364	2.756	3.038	3.659
30	0.683	0.854	1.310	1.697	2.042	2.360	2.750	3.030	3.646
35	0.682	0.852	1.306	1.690	2.030	2.342	2.724	2.996	3.591
40	0.681	0.851	1.303	1.684	2.021	2.329	2.704	2.971	3.551
45	0.680	0.850	1.301	1.680	2.014	2.319	2.690	2.952	3.520
50	0.680	0.849	1.299	1.676	2.008	2.310	2.678	2.937	3.496
55	0.679	0.849	1.297	1.673	2.004	2.304	2.669	2.925	3.476
60	0.679	0.848	1.296	1.671	2.000	2.299	2.660	2.915	3.460
70	0.678	0.847	1.294	1.667	1.994	2.290	2.648	2.899	3.435
80	0.678	0.847	1.293	1.665	1.989	2.284	2.638	2.887	3.416
90	0.678	0.846	1.291	1.662	1.986	2.279	2.631	2.878	3.402
100	0.677	0.846	1.290	1.661	1.982	2.276	2.625	2.871	3.390
120	0.672	0.845	1.289	1.658	1.980	2.270	2.617	2.860	3.373
∞	0.675	0.842	1.282	1.645	1.960	2.241	2.576	2.807	3.291

(Degrees of freedom)

Note » signifies – too large to consider.

Appendix 6: Cumulative Poisson Curves

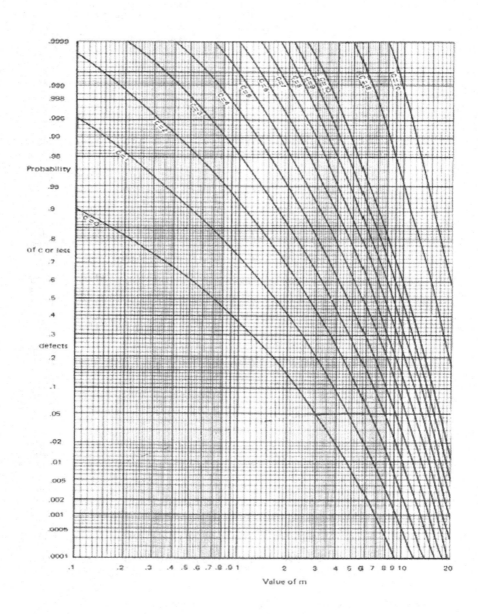

Appendix 7: Percentage Points of the Chi Squared Distribution

These Chi-squared tables are "single-sided" in that they assess the significance of there being a difference in a specified direction.

		Probability that χ^2 > or = to the stated value										
		0.999	0.99	0.9	0.8	0.7	0.6	0.5	0.4	0.3	0.2	0.1	0.01
	1	<0.001	<0.001	0.02	0.06	0.15	0.28	0.46	0.70	1.10	1.64	2.71	6.63
	2	0.00	0.02	0.21	0.45	0.71	1.02	1.39	1.83	2.41	3.22	4.61	9.21
	3	0.02	0.12	0.58	1.00	1.42	1.87	2.37	2.95	3.67	4.64	6.25	11.30
	4	0.09	0.30	1.06	1.65	2.19	2.75	3.36	4.04	4.88	5.99	7.78	13.30
	5	0.21	0.55	1.61	2.34	3.00	3.66	4.35	5.13	6.06	7.29	9.24	15.10
	6	0.38	0.87	2.20	3.07	3.83	4.57	5.35	6.21	7.23	8.56	10.60	16.80
	7	0.60	1.24	2.83	3.82	4.67	5.49	6.35	7.28	8.38	9.80	12.00	18.50
Degrees of freedom	8	0.86	1.65	3.49	4.59	5.53	6.42	7.34	8.35	9.52	11.00	13.40	20.10
	9	1.15	2.09	4.17	5.38	6.39	7.36	8.34	9.41	10.70	12.20	14.70	21.70
	10	1.48	2.56	4.87	6.18	7.27	8.30	9.34	10.50	11.80	13.40	16.00	23.20
	15	3.48	5.23	8.55	10.30	11.70	13.00	14.30	15.70	17.30	19.30	22.30	30.60
	20	5.92	8.26	12.40	14.60	16.30	17.80	19.30	21.00	22.80	25.00	28.40	37.60
	25	8.65	11.50	16.50	18.90	20.90	22.60	24.30	26.10	28.20	30.70	34.40	44.30
	30	11.60	15.00	20.60	23.40	25.50	27.40	29.30	31.30	33.50	36.30	40.30	50.90
	35	14.70	18.50	24.80	27.80	30.20	32.30	34.30	36.50	38.90	41.80	46.10	57.30
	40	17.90	22.20	29.10	32.30	34.90	37.10	39.30	41.60	44.20	47.30	51.80	63.70
	45	21.30	25.90	33.40	36.90	39.60	42.00	44.30	46.80	49.50	52.70	57.50	70.00
	50	24.70	29.70	37.70	41.40	44.30	46.90	49.30	51.90	54.70	58.20	63.20	76.20
	75	42.80	49.50	59.80	64.50	68.10	71.30	74.30	77.50	80.90	85.10	91.10	106.40
	100	61.90	70.10	82.40	87.90	92.10	95.80	99.30	103.00	107.00	112.00	118.50	135.80

Appendix 8: Answers to Exercises

Exercise 1   Manipulating Probabilities

Having dark hair is 80% likely; Having blue eyes is 25% likely; Being bald is 5% likely; Being taller than 1.8 m is 50% likely

1. Having dark hair and blue eyes?
   80% × 25% = 20%
2. Having dark hair or blue eyes?
   Not 80% + 25% = 105% (more than 1!!!!)
   Remember it is Pa+ Pb − PaPb
   So, 80% + 25% − (80% × 25%) = 85%
3. Having dark hair and being bald?
   Not 80% × 5%!!!!
   One cannot be both (they are <u>mutually exclusive</u>)
4. Having dark hair and blue eyes and being taller than 1.8 m?
   80% × 25% × 50% = 10%
5. Having dark hair or blue eyes or being taller than 1.8 m?
   1 − (1−80%) × (1−25%) × (1−50%) = 92.5%
6. Having dark hair and blue eyes or being bald?
   Being bald is mutually exclusive, thus we add the 50% without subtracting PaPb.
   Thus (80% × 25%) + 5% = 25%
7. Having neither blue eyes nor not being bald?
   Not having blue eyes is (1−25%) = 75%
   Not being bald is (1−5%) = 95%
   Thus, having neither is 75% × 95% = 71.25%

Exercise 2 Obtaining the Mean and Standard Deviation

	AGE (a)	NUMBER IN GROUP (b)	TOTAL YRS OF AGE (a)x(b)	MEAN MINUS (a)	THE DIFFERENCE SQUARED	SQUARED DIFFERENCE x (b)	
	6	1	6	2.5	6.25	6.25	
	7	2	14	1.5	2.25	4.5	
	8	4	32	0.5	0.25	1	
	9	4	36	-0.5	0.25	1	
	10	2	20	-1.5	2.25	4.5	
	11	1	11	-2.5	6.25	6.25	
Totals		14	119			23.5	
Mean			8.5			1.68	Variance
						1.30	Std Dev

The totals for the columns no in group, total years, squared difference have used the =SUM(....) Microsoft EXCEL Function. The mean can be obtained either by dividing the 119 total by the cell containing 14 OR by using the =AVERAGE(....) Microsoft EXCEL Function. The variance is the 23.5 cell divided by the 14 cell and the standard deviation by taking the square root (i.e., cell^0.5).

Exercise 3   Mean and Coefficient of Variation

Values of mmol/L

6.90	6.87	6.83	6.90	6.33	6.63	6.67		6.73
0.027778	0.017778	0.01	0.027778	0.16	0.01	0.004444		0.191899
								2.8 %
								Coeff of variation

The mean has been calculated as 6.73.

Below each daily reading is the square of the difference between that value and the mean. By taking the square root of the average, shown in the right-hand column, the standard deviation is obtained.

This is then expressed as a percentage of the mean. Thus, it could be inferred that 68% of the values all fall between + or − 2.8% of the mean value. In other words, the patient's blood sugar levels are significantly consistent.

Exercise 4   Median Life Expectancy

No of Patients	Years of survival	Total Survival Years	Cumulative	Prob of Death	Prob of Survival
2	2	4	4	2%	98%
6	3	18	22	9%	91%
8	4	32	54	21%	79%
11	5	55	109	42%	58%
13	6	78	187	72%	28%
9	7	63	250	97%	3%
1	8	8	258	100%	0%
50		258			

Thus, the median is 5 ½ years and the chance of surviving beyond 6 years is 28%.

Exercise 5   Comparing Two Distributions

The following spreadsheet has been used to calculate the two means and standard deviations and then to calculate the values F and t.

There being 6 items in each sample of data, there are 5 degrees of freedom.

	Group 1	Group 2			
	61	45			
	32	30			
	37	32			
	43	36			
	29	27			
	50	43			
Mean	42	35.5		6.50	Difference of Means
Std Dev	12	7.23		1.14	t = Diff/SqRt(Std²/n+Std²/n)
Variance	144	52.3		2.75	F = Ratio of Variances

From the F tables, the 2.75 is (between 10% and 25%) thus approximately 15% likely to be by chance. In other words, we can only be about 85% sure that the difference is significant and that the Group 2 results were more consistent.

From the t tables the 1.14 is about 30% likely to be by chance. In other words, we can only be about 70% sure that the difference is significant and that the treatment is effective.

Exercise 6 Attributes Using Poisson Curves

A practice covers 2,000 potential patients, and it is known that the incidence of stroke in the UK is approximately 115,000 pa (population 67 million). Thus, the frequency is
= 115k/67 m = 1.7×10^{-3} pa.

Therefore, m = $1.7 \times 10^{-3} \times 2,000 = 3.4$

From the Appendix 6 curves, the probability that the practice will need to treat:

0 cases	= 3%
1 or less case	= 15% thus 1 case 15%–3%=12%
2 or less cases	= 33% thus 2 cases 33–15%=18%
3 or less cases	= 55% thus 3 cases 55%–33%=22%
4 or less cases	= 74% thus 4 cases 74%–55%=19%
5 or less cases	= 87% thus 5 cases 87%–74%=13%

Exercise 7   Chi-Squared

	TREATMENT A WITH INSULIN ONLY					
	BASE DATA	TREATMENT A	ANTICIPATED			
	IMPROVEMENT	IMPROVEMENT	IF NO EFFECT	DIFFERENCE	DIFFERENCE	DIFFERENCE
	> 10% REDUCTION	> 10% REDUCTION	TREATMENT A		SQUARED	SQUARED/ANTICIPATED
	W/O TREATMENT	WITH TREATMENT				
1 MONTH	4	2	1.655172	0.344828	0.118906	0.071839
1-2 MONTHS	15	6	6.206897	-0.2069	0.042806	0.006897
2-3 MONTHS	6	2	2.482759	-0.48276	0.233056	0.09387
>3 MONTHS	4	2	1.655172	0.344828	0.118906	0.071839
	29	12	· 12		ChiSquare =0.244444	
					>95%	likely to have occurred by chance
	TREATMENT B WITH INSULIN & AN ADDITIONAL DRUG					
	BASE DATA	TREATMENT B	ANTICIPATED			
	IMPROVEMENT	IMPROVEMENT	IF NO EFFECT	DIFFERENCE	DIFFERENCE	DIFFERENCE
	> 10% REDUCTION	> 10% REDUCTION	TREATMENT B		SQUARED	SQUARED/ANTICIPATED
	W/O TREATMENT	WITH TREATMENT				
1 MONTH	4	0	2.344828	-2.34483	5.498216	2.344828
1-2 MONTHS	15	8	8.793103	-0.7931	0.629013	0.071535
2-3 MONTHS	6	5	3.517241	1.482759	2.198573	0.625085
>3 MONTHS	4	4	2.344828	1.655172	2.739596	1.168357
	29	17	17		ChiSquare =4.209804	
					40%	likely to have occurred by chance

It might therefore be argued that (**for this group of patients only**) the insulin alone made little or no significant improvement to the condition and that it is just 60% likely that the combined treatment resulted in an improvement. The outcome would therefore be encouraging but not conclusive.

Exercise 8   Correlation

If the spreadsheet has been properly constructed, then the $r = 0.32$ should have been obtained. The formula for the cell D33 is shown below and represents the formula in Chapter 8

=F28/(G28*H28)^0.5

Cells B28–H28 are the sums of the rows 2–25.

Cells B30–C30 are the averages of the rows 2–25.

	A	B	C	D	E	F	G	H
	Item	X	Y	X-Mean	Y-Mean	(X-Mean)x (Y-Mean)	(X-Mean)sq	(Y-Mean)sq
1	Item	X		X-Mean	Y-Mean		(X-Mean)sq	(Y-Mean)sq
2		147	73	-8.3	-0.1	0.7	68.8	0.0
3		156	78	0.7	4.9	3.5	0.5	24.2
4		160	73	4.7	-0.1	-0.4	22.2	0.0
5		189	80	33.7	6.9	233.1	1136.3	47.8
6		173	68	17.7	-5.1	-90.0	313.6	25.8
7		157	70	1.7	-3.1	-5.3	2.9	9.5
8		139	80	-16.3	6.9	-112.7	265.4	47.8
9		157	79	1.7	5.9	10.1	2.9	35.0
10		165	60	9.7	-13.1	-127.0	94.3	171.2
11		156	82	0.7	8.9	6.3	0.5	79.5
12		175	78	19.7	4.9	96.9	388.4	24.2
13		158	70	2.7	-3.1	-8.4	7.3	9.5
14		161	80	5.7	6.9	39.5	32.6	47.8
15		180	74	24.7	0.9	22.6	610.5	0.8
16		145	78	-10.3	4.9	-50.6	105.9	24.2
17		162	69	6.7	-4.1	-27.4	45.0	16.7
18		147	70	-8.3	-3.1	25.6	68.8	9.5
19		168	79	12.7	5.9	75.2	161.5	35.0
20		131	68	-24.3	-5.1	123.5	590.1	25.8
21		138	75	-17.3	1.9	-33.1	299.0	3.7
22		126	58	-29.3	-15.1	441.8	858.0	227.5
23		147	73	-8.3	-0.1	0.7	68.8	0.0
24		145	70	-10.3	-3.1	31.7	105.9	9.5
25		145	69	-10.3	-4.1	42.0	105.9	16.7
26								
27								
28	sum	3727	1754			698.4	5355.0	891.8
29								
30	mean	155.3	73.1					
31								
32								
33			R	0.319591				

The data entered thus indicates a very poor correlation between the systolic and the diastolic. In fact, r^2 becomes about 0.1.

Exercise 9a   Large and Small Numbers

	Write Each as a Number Followed by 10^x	*Write the Answer as a Number Followed by 10^x*
500 groups of 50 people	$5\ 10^2 \times 5\ 10^1$	$25\ 10^3 = 2.5\ 10^4$ people
50 men for 200 years	$5\ 10^1 \times 2\ 10^2$	$10\ 10^3 = 10^4$ manyears
500 bottles divided into 25 crates	$5\ 10^2 / 2.5\ 10^1$	$2\ 10^1$ i.e., 20 per crate
200 deaths from scalding in a popu- lation of 50,000,000	$2\ 10^2 / 5\ 10^7$	$0.4\ 10^{-5} = 4\ 10^{-6}$ chance of scalding
3600 patients awaiting 12 beds	$3.6\ 10^3 / 12$	$0.3\ 10^3$ i.e., 300 patients per bed
1500 influenza fatal- ities in a population of 5,000,000	$1.5\ 10^3 / 5\ 10^6$	$0.3\ 10^{-3}\ 3\ 10^{-4} =$ chance of be- coming a fatality
1 in 500 suffer from diabetes in a popu- lation of 75,000	$(1/500) \times 75\ 10^3$ $= 2\ 10^{-3} \times$ $75\ 10^3$	$150\ 10^0 = 150$ patients*

*Notice how +3 plus −3 = 0 and thus 10^0 which is one.

Exercise 9b   Large and Small Numbers

	Write the Answer as a Number Followed by 10^x	Hints About the Thought Process
$10^6 / 5$	$2 \ 10^5$	5 on the bottom means 2 on the top but reduce the 6 (by 1) to 5.
$6 \ 10^4 / 3 \ 10^2$	$2 \ 10^2$	$6/3 = 2$ and exponent $4 - 2 =$ exponent 2
$5 \ 10^{-3} / 2.5 \ 10^{-1}$	$2 \ 10^{-2}$	$5/2.5 = 2$ and exponent -3 take away -1 (i.e., add 1) gives -2
$9 \ 10^{-1} / 3 \ 10^{-2}$	$3 \ 10^1 = 30$	$9/3 = 3$ and exponent -1 take away -2 (i.e., add 2) gives $+1$
$2 \ 10^{-3} / 4 \ 10^{-2}$	$5 \ 10^{-2}$	$2/4$ means 5 on the top BUT reduce the exponent by 1. Thus, $-3 - 1 - (-2) = -2$
$10^{-4} / 5 \ 10^{-5}$	$2 \ 10^0 = 2$	5 on the bottom means 2 on the top but reduce the exponent by 1. Thus, $-4 - 1 - (-5) = 0$
$3 \ 10^9 / 6 \ 10^7$	$5 \ 10^1 = 50$	$3/6 = 5$ but reduce the exponent by 1. Thus, $9 - 7 = 2$. Then $2 - 1 = 1$

Exercise 10   SLY/Cost Per Life Saved

The cost per patient screened, given the 50% take-up, is $2 \times £150 = £300$.

It could be argued that 67% of 10% of recipients enjoy an additional 2 years of life.

Thus, the SLY based on the 2 years additional life expectancy:

$$£300/(0.67 \times 0.1 \times 2) = £2,240$$

Given the above assumptions, against a criterion of £60,000 per SLY it might be argued that the programme is highly cost effective.

The reader may care to research these assumptions and construct his/her own argument.

Appendix 9: Further Reading

1. Facts from Figures by M J Moroney (ISBN: 9780140202366).
2. Statistical Methods, 8th edition William G. Cochran George W. Snedecor ISBN 978-8126551385.
3. Technis Guidelines T810 Fatal accident frequency statistics FAFRs – A summary, from david.smith@technis.org.uk
4. Reliability, Maintainability and Risk, 9th Edition (2017) Smith DJ, (Elsevier) ISBN 978008102010-4.

Index

Printed in the United States
by Baker & Taylor Publisher Services

Printed in the United States
by Baker & Taylor Publisher Services